T0325682

Power Electronics Applied to Industrial

Systems and Transports 1

Series Editor
Bernard Multon

Power Electronics Applied to Industrial Systems and Transports

Volume 1
Synthetic Methodology to Converters and Components Technology

Nicolas Patin

ELSEVIER

First published 2015 in Great Britain and the United States by ISTE Press Ltd and Elsevier Ltd

ISTE Press Ltd
27-37 St George's Road
London SW19 4EU
UK

www.iste.co.uk

Elsevier Ltd
The Boulevard, Langford Lane
Kidlington, Oxford, OX5 1GB
UK

www.elsevier.com

Notices
Knowledge and best practice in this field are constantly changing. As new research and experience broaden our understanding, changes in research methods, professional practices, or medical treatment may become necessary.

Practitioners and researchers must always rely on their own experience and knowledge in evaluating and using any information, methods, compounds, or experiments described herein. In using such information or methods they should be mindful of their own safety and the safety of others, including parties for whom they have a professional responsibility.

To the fullest extent of the law, neither the Publisher nor the authors, contributors, or editors, assume any liability for any injury and/or damage to persons or property as a matter of products liability, negligence or otherwise, or from any use or operation of any methods, products, instructions, or ideas contained in the material herein.

For information on all Elsevier publications visit our website at
http://store.elsevier.com/

British Library Cataloguing in Publication Data
A CIP record for this book is available from the British Library
Library of Congress Cataloging in Publication Data
A catalog record for this book is available from the Library of Congress
ISBN 978-1-78548-000-3

Printed and bound in the UK and US

Contents

Preface

The purpose of this series of books is to give an in-depth presentation of the field of power electronics. It will be split into four volumes, comprising a total of 21 chapters (plus appendices).

Volume 1 is aimed at introducing essential notions in power electronics from both theoretical and technological perspectives, with a focus on source connection rules, reversibility and the impact on the choice of switches for converter synthesis. Concerning technological aspects, the standard active components will be presented from a "user" perspective, and include certain elements regarding their packaging and thermal characteristics, with electrical modelings which are equivalent in terms of both permanent and transient modes. This aspect will be presented from a "user" perspective with a focus on the selection and dimensioning of cooling equipment. We will also consider component control (transistors, thyristors, triacs, etc.) and snubbers. The local environment of electronic switches will therefore be covered in some detail. A separate chapter is devoted to passive components (capacitors and magnetic components – inductors and transformers), highlighting the criteria involved in the choice of capacitors based on the technologies and limitations of each option. We will provide a more detailed discussion of magnetic components, with a guide to the choice of magnetic circuits (materials and layout) and the dimensions of windings (number of turns, use of Litz wires, etc.). The final chapter will consider the elements involved in the design and production of printed circuits.

Volume 2 [PAT 15a] will deal with industrial applications, notably those linked to transporting electronic power converters, with a focus on power supplies for electrical machinery. We will consider different types of DC/DC, DC/AC, AC/DC and AC/AC converters. We will also provide an introduction to multi-level converters (primarily in the context of inverters). The final chapter of this volume presents a case study involving full dimensioning of an industrial variable-speed drive.

Volume 3 [PAT 15b] is concerned with converters (essentially of the DC/DC variety) in the context of switch-mode power supplies, another key area in which power electronics is used is in the supply of energy to a variety of electronic equipment for signal and information processing. We will also discuss the dimensioning of inductors and HF transformers. Volume 3 also provides an introduction to soft switching, using the specific example of a DC/DC converter using a resonant inverter. A further chapter in this volume is devoted to modeling in the context of controlling switch-mode power supplies, based on Middlebrook's equivalent average models. Finally, as in Volume 2, a case study will be used to provide a thorough overview of the design of a digitally-controlled Flyback power supply, used for practical exercises at the UTC.

Volume 4, [PAT 15c] the last volume in the series, is devoted to electromagnetic compatibility, and is divided into three chapters. In Chapter 1, we will introduce and discuss the modeling of common sources of disturbances before considering case studies in Chapters 2 and 3 in order to gain an understanding of the pathways taken by these disturbances to reach their targets. This chapter structure does not fully conform to the usual notions of conducted and radiated disturbances, but distinguishes between "circuit" couplings (i.e. localized constant models – Chapter 2) and those involving propagation phenomena (i.e. distributed constant models – Chapter 3).

The case study chapters should not be considered simply as applications of the theoretical concepts developed in other chapters: certain concepts (both technological and practical) are introduced using these examples, in situations where these specific cases are sufficiently representative of a wide range of applications.

Finally, three appendices cover the usual formulas used in power electronics and electrical engineering (Appendix 1, see Volumes 1–4), a detailed presentation of spectrum analysis for periodic and non-periodic signals (Appendix 2, see Volumes 2 and 4) and a compilation of technical documentation for the components discussed elsewhere in the book (Appendix 3, see Volume 3).

This book forms the basis for power electronics teaching at the *Université de Technologie de Compiègne* (UTC, or Compiègne Technological University), in France. The first of the two modules on offer exclusively covers power electronics, including all of the families of converters presented in Volume 2. The main focus of this module is on power supplies for electrical apparatus, both in industry and for transportation. The module is offered exclusively to students in the Mechanical Engineering Department at the UTC, specifically to those following the Mechatronics, Actuators, Robotics and Systems (MARS) course. Switch-mode power supplies (Volume 3) are covered as part of a general module on electronic functions for engineering, with a lecture, seminar and practical session devoted to the subject, based on a flyback-type power supply. This more general module has a wider audience and is also offered to students in biological engineering with a specialization in biomedical sciences. This series of books therefore takes a broad approach to the elements covered in the second module, covering both the different converter topologies used in designing switch-mode power supplies and the design methodology involved (notably through a case study). We also cover certain aspects of control, more specifically control-based modeling, in order to produce usable models using automatic methods to regulate the output voltage independently of variations in load consumption. EMC (Volume 4) is also covered as part of this more general module, but is only touched on briefly.

Nicolas PATIN
Compiègne, France
December 2014

Introduction

I.1. Generic structure of an industrial variable-speed drive

A variable-speed drive such as the one shown in Figure I.1 is a useful subject for study as it allows us to include most of the functions involved in power electronics; furthermore, devices of this type are widely used in an industrial context. Factories are generally powered by a fixed-frequency alternating network (230–400 V/50 Hz in France), and most of the electrical machines (robots, conveyors, machine tools, etc.), which are used in production lines, with a power requirement of greater than or equal to 1 kW (the most widespread), use a three-phase alternating current (AC) (these are usually induction machines with squirrel cage rotors, or, more rarely, permanent magnet synchronous machines).

In the cases where a variable speed function is required, these machines require a power supply with a variable frequency (and amplitude). However, the electrical network provides a fixed frequency. As we will see, it is easier to produce voltages of variable frequency (and amplitude) using a continuous source (using an *inverter*) than to transform an AC of frequency f_1 into a new AC with frequency $f_2 \neq f_1$ (the device used for this purpose is known as a *cycloconverter* and will be presented in Chapter 1, but will not be studied in detail as they are no longer widely used). The following chapters will highlight the *modular nature of electronic power converters*, as in the case of the variable speed-drive shown in Figure I.1. This device uses a *rectifier*, which allows us to pass from the *AC with fixed amplitude and frequency* provided by the power network to the continuous supply required by the *inverter* to produce a network with an *AC with variable amplitude and frequency* for the electrical device in question. The insulated gate bipolar transistors (IGBT) inverter acts as a rectifier (during machine braking phases), but the diode rectifier used here is unable to return this electrical energy to the network. This means that a *brake chopper* is needed to consume this excess energy. These different converters will be covered, in turn, over the course of the following chapters.

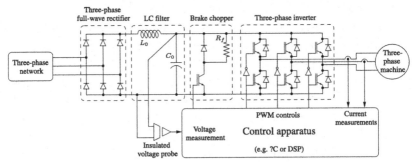

Figure I.1. *"Typical" power design of an industrial variable-speed drive*

Before beginning our study of different converter layouts, it is important to consider the specificities of power electronics. This domain involves the control of electrical power, from the power supply to the powered load. The uses of this function are evident, but the methods used need to be covered in more detail.

I.2. Specificities of power electronics

I.2.1. *Distinction between signals and energy*

Unlike the branch of electronics that concerns *information processing* equipment (whether analog or digital, using operational amplifiers, A/D or D/A converters, microprocessors, microcontrollers or other programmable or non-programmable digital circuits), power electronics focuses on *constructions that process and transform electrical energy to respond to a need*. Nevertheless, there are certain similarities between the processing of digital information (digital electronics) and power electronics, due to the way in which electronic components function in switching operations. In power electronics, transistors (among other components) are used in turned off/saturated mode, which renders them equivalent to open or closed switches. While these modes of operation are similar, they do not however concern the same objectives. In digital electronics, this type of function serves multiple purposes. Examples include:

– noise immunity in digital circuits;

– programming new functions in an unmodified physical circuit[1];

– improved integration (smaller circuits).

In power electronics, however, the sole aim of using switch functions in components is to *maximize efficiency* (and therefore *minimize loss*) in the converter.

I.2.2. *Commutations and losses*

The interest of switching-based functions may be observed by studying the (simple) construction shown in Figure I.2.

---

1 Note that programmable analog circuits do exist, but these too use switches – see *switched capacitor filters* [GHA 03].

This construction is made up of an input voltage source E (e.g. a battery) connected to an association in series with a resistance R (modeling, e.g. a lamp, a heating element, etc.) and a transistor T (in this case, an NPN bipolar transistor).

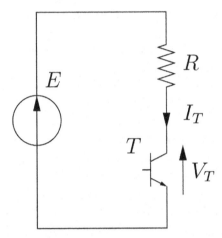

Figure I.2. *"Simple" power modulation circuit*

The equation of this simple single-loop circuit is easy to establish:

$$E = R.I_T + V_T \qquad\qquad [I.1]$$

The operating point of this circuit is (graphically) the intersection between the characteristic $V_T(I_T)$ of the transistor T and the load line $V_T = E - R.I_T$ produced by equation [I.1], as shown in Figure I.3. Note that the characteristic of the transistor is linked to the command signal applied at the base. For example, the operating point M corresponds to a base/emitter voltage $V_{BE} = V_{BE5}$: we thus obtain a voltage of V_{T0} between the collector and the emitter of the transistor, and the current I_T traveling through the transistor is equal to I_{T0}. The power dissipation of any dipole is the product of voltage $\times$ current, in this case $V_T.I_T$. The temperature of a component is directly linked to the power dissipation, and constructor documentation for electronic components specifies a maximum junction temperature for operation without damaging the component. Consequently, an electronic component (associated with correctly dimensioned cooling equipment) may be characterized by a maximum power $P_{\max}$. The *isopower curve* $P_T = V_T.I_T = P_{\max}$ is shown in dotted lines in Figure I.3. We see that point M lies above this curve, and cannot therefore be accessed in steady state. In fact, use of the transistor in the linear zone leads to significant overdimensioning in relation to the power required by the load. This is generally the case for low-power requirements (e.g. class A audio amplifiers) but is not possible for high powers when a high yield is required, notably for equipment with integrated energy, where priority is given to autonomy and/or compactness (using smaller, lighter thermal dissipation elements).

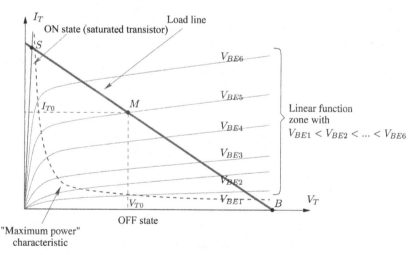

Figure I.3. *Operating point of the circuit*

However, the use of the "turned off" and "saturated" transistor modes alone seems to be limiting in terms of the dosage of the power supplied to the load, in which we must either provide no power, in cases where the transistor is operating as an open switch (point B), or maximum power, practically equal to E^2/R, if the transistor is saturated (closed switch – point S).

I.2.3. Load inertia and average model

The disadvantage presented by the ON/OFF operating mode for transistors, as discussed above, does not pose significant problems in practice for most applications, due to the *inertia of the powered load*. The transistor is assimilated to a switch, but, unlike in electromechanical models, this switch may operate using very short open/closed cycles (duration denoted as T_d, the *switching period*) without incurring damage. It is thus possible to make cycles sufficiently short for the powered load to be perfectly insensitive to commutation; it therefore behaves in a manner that is *equivalent to a linear power supply with a voltage equal to the average voltage supplied by the switching circuit*. The period T_d must be short in relation to the time constant of the load:

– thermal time constant for a heating element (generally high – from a number of seconds to a number of minutes in the case of ovens);

– thermal time constant, again, for the filament of an incandescent light bulb (tens or hundredths of a millisecond);

– biological time constant of persistence of vision for a rapid light source[2] such as white LEDs[3]($T_d < 40$ ms).

An LED powered with a nominal voltage/current 50% of the time will produce a light flow equal to 50% of the nominal flow.

This very general principle also applies to electrical machines, and progress in the design of electronic

---

2 This is necessary in order to avoid flickering that may be uncomfortable for the user of the light source.
3 LED = light emitting diode

components means that we can now attain frequencies of the order of around 10 kHz for industrial variable-speed drives with power levels greater than or equal to 1 kW. Evidently, switches commutating low-power levels can reach higher switching frequencies than high-power switches: these components themselves are subject to the same inertia problems as the loads they power. Note, for example, that switching frequencies are often lower than 1 kHz for motors with power levels measured in megawatts, such as those used in railway engines.

I.3. Families of converters

I.3.1. *Classification of structures*

The most obvious classification system for a first analysis of electronic power converters consists of differentiating between contexts of use. Sources, in the broadest sense (including loads), which are interconnected by converters, may be split into two main families:

– direct current (DC) sources;

– AC sources.

Consequently, when imagining a converter involving two different sources, we can identify $2^2 = 4$ distinct types (plus a fifth type due to the two available types of AC/AC converter):

– DC/DC converter, known as a *chopper*;

– DC/AC converter, known as an *inverter*;

– AC/DC converter, known as a *rectifier*;

– AC/AC converters, known as *dimmers* if the input and output frequencies are the same, or *cycloconverters* if this is not the case.

These different configurations are shown in Figure I.4.

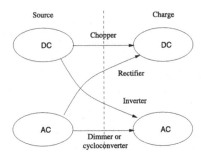

Figure I.4. *Converter structures*

I.3.2. *Typical waveforms*

Before going into detail analyzing the structures of electronic power converters, we should illustrate their operation using the waveforms of the voltages they may supply to a load in order to show their ability to modulate the transmitted energy. Figure I.5 shows the possible output waveforms of a chopper, while Figure I.6 shows the voltage waveforms produced by an inverter. Figures I.7 show the waveforms generated by dimmers commanded, respectively, by "phase angles" (e.g. for lighting) and "wave trains" (for high-power levels, e.g. for heating).

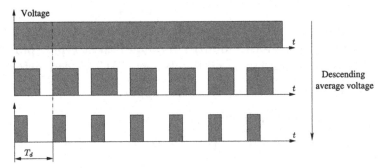

Figure I.5. *Output voltage waveforms from a chopper*

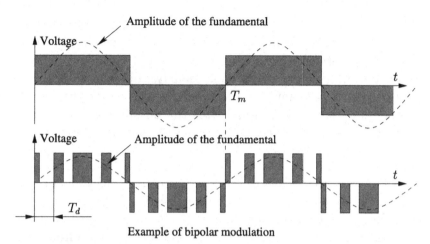

Example of bipolar modulation

Figure I.6. *Output voltage waveforms from an inverter*

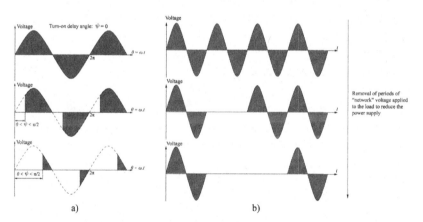

Figure I.7. *Output voltage waveforms from a dimmer using two different control types: a) phase angle and b) wave train*

I.3.3. *Orders of magnitude and applications*

The electrical power levels handled by electronic power converters cover several orders of magnitude, from converters with a nominal power of approximately 1 W to converters dealing with powers of over 1 GW at the other extreme (e.g. the High Voltage Direct Current (HVDC) cross-channel DC connection between France and the United Kingdom, which has a nominal power of 2 GW). Clearly, this broad range of power values cannot be handled by the same types of components; however, it illustrates the possibilities currently offered by semiconductor components, which continue to evolve even as we speak.

It is useful to discuss orders of magnitude and widespread applications of power electronics before going on to consider electronic switch technologies, characteristics and the structures of converters. We will begin by defining a number of standard voltage values used in three main families of electric power supplies:

– DC voltages;

– single-phase AC voltages;

– three-phase AC voltages.

These values are shown in Table I.1 alongside their corresponding general applications. Note, however, that this list is in no way intended to be exhaustive.

Volume 2 [PAT 15a] in this series will focus on the use of electronic power converters in supplying electrical machinery. However, the following list, along with Figure I.8, shows a wide variety of possible applications:

– DC: robotics, electronic apparatus (i.e. switch-mode power supplies), industrial electrolysis, soldering and battery chargers;

– single-phase AC: domestic heating, lighting, uninterrupted power supply and rectifiers for rail traction on AC networks;

– three-phase AC: rail traction motors, electric or hybrid vehicles, medium- and high-powered industrial actuators and stabilization of electrical networks (Static Synchronous Compensator (STATCOM), Flexible AC Transmission System (FACTS), etc.).

Switch-mode power supplies constitute an important field of application of power electronics, and will be discussed in detail in Volume 3 [PAT 15b].

Format	Voltage (V)	Applications
DC	1.5; 3; 4.5; 9; etc.	Power supply for small portable devices
DC	12	Car battery (onboard automobile network)
DC	750	Subway
DC	1,500; 3,000	Rail traction on certain networks
AC 1	230 V / 50 Hz	European domestic network
AC 1	115 V/60 Hz	US domestic network
AC 1	25k V/50 Hz	Rail traction on certain networks
AC 1	15 kV/16 Hz2/3	Rail traction in Switzerland and Germany
AC 3	230–400 V/50 Hz	European domestic network
AC 3	115–200 V/60 Hz	US domestic network
AC 3	20 kV/50 Hz	French distribution network
AC 3	115–200 V/400 Hz	Onboard network in aircraft (currently under development)

Table I.1. *Voltage levels and applications*

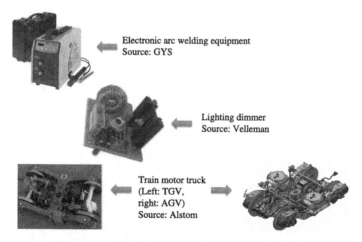

Electronic arc welding equipment
Source: GYS

Lighting dimmer
Source: Velleman

Train motor truck
(Left: TGV,
right: AGV)
Source: Alstom

Figure I.8. *Applications of power electronics*

Theoretical Tools and Active Components for Power Electronics

1.1. Electrical circuits and power electronics

1.1.1. *General case*

In studying *linear electric circuits*, we distinguish between classic passive dipoles (which absorb energy, although this energy value may be null):

– *resistors;*

– *inductors;*

– *capacitors.*

And active dipoles (which supply energy: this is generally positive, but can also be negative) in the form of *sources* of:

– *voltage* (voltage at the terminals is imposed, and independent of the current delivered);

– *current* (carrying an imposed current, independent of the voltage at the terminals).

Note that, generally speaking, a dipole consisting of a current source and another dipole will behave in the same manner as the current source alone in relation to the circuit to which it is connected. At the same time, we see that a dipole consisting of a voltage source in parallel with another dipole behaves in the same way as the voltage source alone in relation to the circuit.

Another useful point to note relates to the interconnection of sources. The connection of two current sources creates a conflict between the two elements, as they may impose different values on the same current. Similarly, the parallel use of two voltage sources creates conflict if their values are different. However, it is entirely possible to connect two sources of different natures in series or in parallel (i.e. a voltage source with a current source).

These definitions are used "as is" or with minor modifications in the context of power electronics, as we will see later in this chapter.

1.1.2. *Extension to power electronics*

These components are all used in power electronics, but the definitions of voltage and current sources are less restrictive in this domain than in more general situations:

– in power electronics, a voltage source is a dipole with terminals where the voltage cannot be discontinuous;

– similarly, a current source is a dipole carrying a current which cannot be discontinuous.

Thus, the sources defined in a general context are still considered in the same manner in power electronics, but, by extension, we consider that:

– a dipole obtained by connecting an inductor and any dipole in series is a current source;

– a dipole obtained by connecting a capacitor and any dipole in parallel is a voltage source.

As we have already seen, power electronics is not limited to the use of passive linear dipoles (resistors, inductors and capacitors) and voltage or current sources, but makes use of switches (transistors, diodes, etc.). Consequently, we need to open circuits (OFF state) or provoke short-circuits (ON state):

– an open circuit may be assimilated to a source of null current;

– a short-circuit may be assimilated to a source of null voltage.

The rules defined above concerning the interconnection of voltage and current sources may be applied to this basis, enabling us to:

– short-circuit a current source;

– open the circuit of a voltage source.

But we cannot:

– short-circuit a voltage source;

– open the circuit of a current source.

This final point is important in the construction shown in Figure 1.15, where a free-wheeling diode is needed in parallel to an inductive load (i.e. a current source in the sense of power electronics). This set of rules is presented in Figure 1.1. A classification of sources/loads according to their nature (current source/voltage source) is given in Table 1.1 (this is not an exhaustive list).

The ideal switch K (see Figure 1.2) is a dipole which may be considered to have infinite resistance when open, and zero resistance when closed. It can therefore carry a positive or negative current I_K when open and tolerate a positive or negative voltage V_K at its terminals when turned off. This type of component is *bidirectional for both voltage and current*. This gives us a voltage/current characteristic made up of two full lines following the two axes of the plane (I_K, V_K). Each half-line of this characteristic, starting from the origin of the plane, is known as a *segment*. The ideal switch is thus a *4 segment switch*. As we will see, in practice, real components have a lower number of segments (with the exception of the triac).

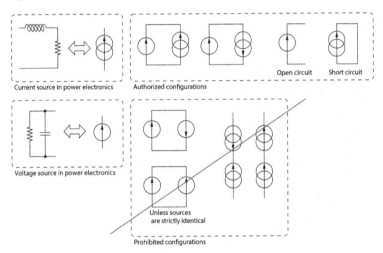

Figure 1.1. *Rules for circuits in power electronics*

Current sources	Voltage sources
DC machine	Battery
Synchronous machine (generally three-phase)	Solar panel
Asynchronous machine (generally three-phase)	Fuel cell
Transformer winding	Piezoelectric transformer
Inductor dipole in series	Condenser dipole in parallel
Electrical network with finite power	Electrical network with infinite power

Table 1.1. *Classification of current or voltage source types*

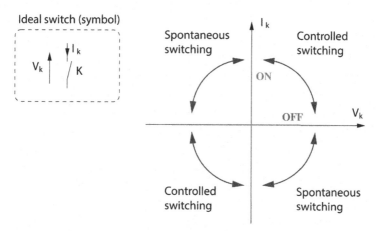

Figure 1.2. *The ideal switch and its current / voltage characteristic*

The open (turned off – noted OFF) and closed (noted ON) states of the switch are known as *static states*. The switch must also be characterized dynamically by the possible transitions between ON and OFF segments. The transition from an OFF segment to an ON segment is known as *turn-on*; the opposite transition is known as *turn-off*. These two switching types are themselves split into subtypes according to the sign of the starting segment and the sign of the destination segment:

– switching between two segments with the same sign is *controlled* (and dissipative);

– switching between two segments with opposite signs is said to be *spontaneous* (and non-dissipative).

This is clearly shown in the diagrams in Figure 1.3. The switch cannot produce energy, and as the *convention* used for the voltage V_K and the current I_K is that of a *receiver*, the switching trajectories cannot pass through quadrants 2 and 4 (but only through quadrants 1 and 3). This clearly shows why controlled switching is dissipative (and should therefore be carried out as quickly as possible to minimize *switching losses*), while the trajectory is forced to follow the axes (implying an instantaneous power value of zero) in the case of spontaneous switching.

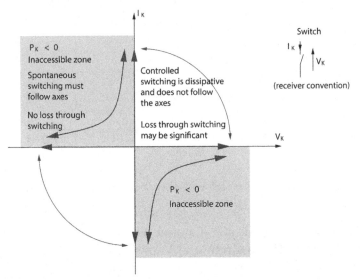

Figure 1.3. *Controlled and spontaneous switching*

1.2. Real components

1.2.1. *Two-segment switches*

1.2.1.1. *Diodes*

A diode is a component made up of a doped semiconductor (Si) crystal (zone with a majority of holes), P, at one extremity, known as the *anode*, and a region with a majority of electrons, N, at the other extremity, known as the *cathode*. The PN junction has the property of allowing current to pass in one direction only (from the anode to the cathode). The generic symbol for a diode is shown in Figure 1.4 alongside the idealized characteristic $I_d(V_d)$. Diodes are non-controlled components, where the establishment of conduction and turn-off occur *spontaneously*:

– turn-on occurs when the voltage V_d increases to reach a threshold value $V_{\text{threshold}}$ ($V_{\text{threshold}} = 0$ in the idealized model). The diode is clearly turned off ($I_d = 0$) when $V_d < V_{\text{threshold}}$. Once conduction is established, we obtain $V_d = V_{\text{threshold}}$;

– turn-off occurs when the current I_d cancels itself out ($I_d > 0$, initially). The current thus becomes identically null.

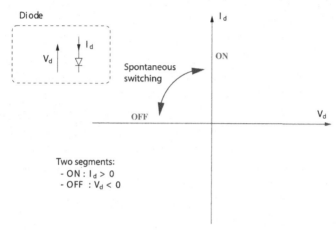

Figure 1.4. *Generic diode and its current/voltage characteristic*

While spontaneous switching is, in theory, non-dissipative, diodes are not actually able to instantly re-establish a voltage at terminals when the current cancels out. At the interface between the P and N zones, the majority hole and electron zones on either side of the border have a tendency to recombine. Through recombination, these zones, initially neutral in electrical terms, acquire space charges. In this zone, an electrical field appears, and is reinforced when we apply an inverse voltage to the diode: in this way, the diode is able to oppose the establishment of a current. This phenomenon is responsible for the OFF behavior of the diode. When the diode is in a state of conduction, this intrinsic field (the cause of the threshold voltage of the diode) is overcome and a current is established. However, part of the charges circulating at the moment conduction is established do not contribute to the current flowing through the diode, but instead *neutralize the depletion region*. When the diode turns off following this conduction phase, the current must not only cancel out, but also invert for a short period in order to evacuate the stored (similar to neutralization) charges. This is the reverse recovery phenomenon. The current then follows a waveform analogous to that shown in Figure 1.5.

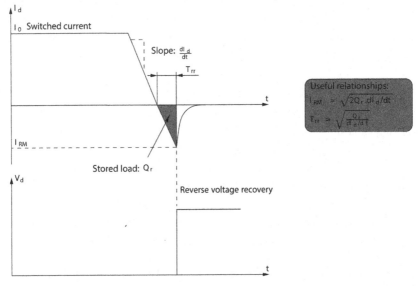

Figure 1.5. *Diode turn-off – reverse recovery*

This phenomenon is important when dimensioning a converter, as the maximum value of the inverse current is a function of two parameters which must be taken into account by the designer:

– the load Q_{RR} stored in the diode, which depends on the technology used and is determined by the manufacturer;

– the speed in reduction of the current, which depends on the environment of the diode (essentially on the piloting of a transistor associated with the diode in a diode half-bridge, as we will see).

These points will be discussed in further detail in Chapter 2, section 2.1.2, and show that the choice of turn-on speed for controlled switches involves finding a compromise between tolerable losses in components and the limitations of the diodes in question.

1.2.1.2. Transistors

1.2.1.2.1. Bipolar transistors

A bipolar transistor is a superposition of three alternately doped semiconducting zones, known as the collector (C), base (B) and emitter (E). The transistor shown on the left in Figure 1.6 is an NPN transistor, while the right hand side of the diagram shows a PNP transistor. The main difference between these two transistor types lies in the way in which they are controlled. As they are considered to be obsolete, we will only consider NPN transistors in this section. Note, however, that in both cases, based on the presentation given in the previous section, the two types of transistor are simply a serial association of two inverted diodes. This representation does not allow a clear understanding of the way in which these transistors operate. The physics involved in semiconductors lies outside of the framework of this book; readers interested by this subject may consult [MAT 09], which provides a detailed study of the physical phenomena at work in these components, among others. The main difference between a transistor and two diodes in series lies in the thickness of the intermediate layer (base), which is very thin in transistors, allowing minority carriers (electrons, in the case of an NPN transistor) to cross without undergoing recombination (majority carriers, i.e. holes here).

From a user perspective, bipolar transistors behave by amplifying the base current I_B at the level of the collector current I_C by a coefficient h_{FE}: $I_C = h_{FE}.I_B$. Coefficient h_{FE} is highly variable from one transistor to another, but for our purposes, we may consider it to be equal to 100 for low-power transistors, but with the ability to take values of less than 10 for high-power transistors. This low gain is highly problematic, and this is why two transistors are cascaded (as shown in Figure 1.7, with a global current gain of the order h_{FE}^2) in a so-called Darlington structure. This technology is no longer used, as the control of a conducting transistor consumes current, a fact which inevitably reduces system yield and adds complexity when designing control organs (preamplifiers). However, the technology has left a visible legacy in Insulated Gate Bipolar Transistors (IGBT) technology, which will be presented later; bipolar translators perform well at high voltages. This is not the case with Metal Oxide Semiconductor Field Effect Transistor (MOSFET) technology, presented in the next section, which is better suited for low voltages.

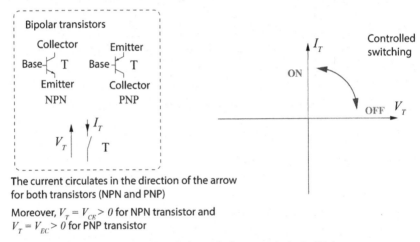

The current circulates in the direction of the arrow for both transistors (NPN and PNP)

Moreover, $V_T = V_{CE} > 0$ for NPN transistor and $V_T = V_{EC} > 0$ for PNP transistor

Figure 1.6. *Symbols and characteristic $I_T(V_T)$ of NPN and PNP transistors*

1.2.1.2.2. MOSFET field effect transistors

The operating principle of the MOSFET is fundamentally different than that of a bipolar transistor. The structure and symbol of an N-channel MOSFET are shown in Figure 1.8.

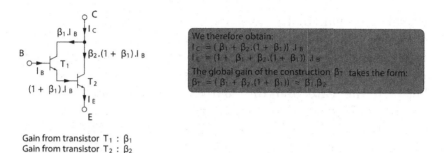

We therefore obtain:
$$I_C = (\beta_1 + \beta_2.(1 + \beta_1)).I_B$$
$$I_E = (1 + \beta_1 + \beta_2.(1 + \beta_1)).I_B$$
The global gain of the construction β_T takes the form:
$$\beta_T = (\beta_1 + \beta_2.(1 + \beta_1)) \approx \beta_1.\beta_2$$

Gain from transistor T_1 : β_1
Gain from transistor T_2 : β_2

Figure 1.7. *Illustration of a Darlington structure*

Structure of an N-channel MOSFET

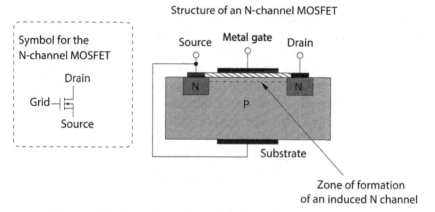

Figure 1.8. *Structure and symbol of an N-channel MOSFET*

As we see from the illustration, the transistor consists of a metallic gate (G) separated from a semiconductor crystal by a layer of isolating oxide (hence the MOS acronym used to denote this structure). The metal gate is placed opposite to a zone P, delimited by two N doped "dots", supporting metallic connections known as the drain (D) and source (S). The structure is perfectly symmetrical, except for the fact that the source is connected to the substrate (base part of the semiconductor crystal). As we have already seen, zone P is a region with a majority of holes. However, electrons (which are free) are still present, as (at any temperature above 0 K) electron–hole pairs are generated systematically due to thermal motion. In the absence of an electrical field, these electrons follow erratic trajectories with a null average shift before recombining with the majority holes. However, in the presence of an electrical field (generated by a voltage V_{GS} between the gate and the source – i.e. the substrate), as shown in Figure 1.8, the electrons are attracted to the insulating layer, while the holes are repelled. This phenomenon results in a reversal of the proportions of electrons and holes in a thin layer of the semiconductor: in the immediate vicinity of the insulating oxide, an N *channel* is created by the field, connecting the N dots, the configuration of which remains unchanged. Current is therefore able to circulate between the drain and the source.

In the context of our applications, in the ON state, a MOSFET transistor behaves in the same way as a conductor with resistance R_{DSon} defined by the manufacturer, and, in the OFF state, as an ideal open circuit. The main interest of these components lies in the fact that the gate is insulated, and no current is required to maintain conduction in the transistor (non-dissipative control). The control circuit is therefore simplified in comparison with that required by a bipolar transistor. However, the control process does consume energy during switching (potentially high quantities of energy in cases of frequent switching), as the gate behaves as a capacitor and the capacity between the gate and the source can be very high for certain types of MOSFET (e.g. the trench gate MOSFET, or Trench MOS, for applications with very low voltages).

The main drawback of the MOSFET compared to bipolar transistors is seen in cases involving high voltages (> 600 V[1]) as the length L_c of the channel is directly linked to the desired withstand voltage (in the "OFF" state). The

---

1 This value continues to evolve due to progress in semiconductor technologies.

resistance in the ON state, as for all conductors, takes the form $\rho\frac{L_C}{S_C}$, where ρ is the conductivity of the material. As the rated voltage increases, the resistance R_{DSon} increases and becomes less favorable compared to that observed in bipolar transistors (where the voltage drop is essentially of type $V_{CEsat} = cte$, with a dynamic resistance which remains low for transistors of all ratings) above a certain power level.

Nevertheless, MOSFET transistors remain interesting from a control standpoint, and work has been carried out to develop a hybrid component for medium voltages (600–1,700 V), combining the advantages of bipolar transistors for the "power" aspect and of MOSFET transistors for control purposes.

1.2.1.2.3. IGBT transistors

IGBTs may be considered as a fusion of the advantages offered by bipolar and MOSFET technologies. This fusion makes use of the central idea of the Darlington structure, which was initially developed for bipolar transistors (see Figure 1.9).

Symbol for an IGBY   Simplified representation of the internal structure

Figure 1.9. *Presentation of the IGBT transistor*

Currently, MOSFET and IGBT technologies are perfectly complementary: a fuzzy threshold between these types can be localized around 600 V.

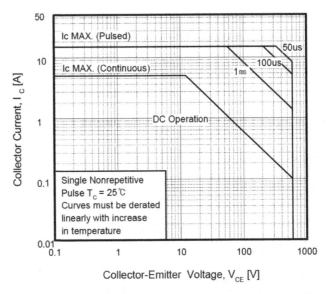

Figure 1.10. *SOA of an IGBT (extract from a Fairchild application note – AN-9016)*

Below this value, MOSFET technology is dominant, for example for powering small motors (model vehicles, small actuators for onboard applications, automobile purposes, switch-mode power supplies for IT, domestic applications, etc.). Above the threshold, IGBT technology is dominant in industrial applications, such as speed variation in synchronous and asynchronous machines, and rail traction, where it is increasingly used to replace components such as Gate Turn OFF (GTO) thyristors, now limited to uses involving very high voltages. These latter components, which will be presented in the next section, are becoming increasingly rare due to the difficulty of use compared to the relative simplicity offered by IGBTs, particularly in terms of safe operating areas (SOAs) as shown in Figure 1.10. The figure is taken from a Fairchild application note: an IGBT, switched in a matter of microseconds, presents a safety operating area which is practically rectangular in the plane (I_K, V_K). This means that, whatever the load type involved, if the rated voltage and current are chosen with care and the component is correctly cooled, switching assistance circuits (see the following chapter) should not be required to guarantee the safe operation of the converter.

Let us consider this figure in greater detail. The curves show that the maximum permissible current for this particular IGBT is between 10 and 20 A. Above this value, the component will be destroyed almost instantly. In the same way, the rated voltage is clearly shown, and must not go above 600 V for all of the defined zones. Note that this result can be generalized to all components: the voltage limit is strict and is independent of the duration of application. Components destroyed by excess voltage are subject to a sudden (and extremely rapid) burnout phenomenon, without overheating, unlike cases of overloading (or excess intensity). This is the reason two current limits are shown in the figure: the first is noted Ic MAX. (Continuous) (between 5 and 6 A), indicating that this current can be permanently handled by the component if the component is cooled correctly (N.B.: this is not necessarily always easy to achieve in a reasonable manner in practice. Care is needed when considering data provided by manufacturers and careful consideration should be given to technical characteristics). Permanent operation is not possible above this value, but pulses are permitted (hence the term "Ic MAX. (Pulsed)"). The oblique limits of the operating zones represent a power limit (and consequently the limit of thermal dissipation). This limit differs between permanent and pulsed modes, as these different cases involve the notion of the thermal time constant of the component chip (see section 2.4): these limits guarantee that internal overheating will not occur in all usage situations. The indicated durations are particularly long (50 μs is the shortest duration given); IGBT switching generally takes less than 1 s. We may therefore expect the authorized zone to be perfectly rectangular. Note, however, that these ultra-short pulses are subject to rapid repetition (thousands or tens of thousands of times per second, depending on the switching frequency of the converter); this takes us outside of the context of non-repetitive pulses, which is the case used for the lines shown in Figure 1.10. We therefore need to study switching losses in order to compare them to the authorized loss budget in continuous mode (100 W according to the limit line for the DC Operation zone), minus conduction losses (in saturated mode) for the target application.

1.2.2. *Three-segment switches*

1.2.2.1. *Thyristors*

Thyristors have been around for a long time (becoming commercially available in 1956) and are well-understood in the context of high-powered applications. Their main use is in controlled rectifiers. Thyristors are slow components, but perfectly suited to use in 50 Hz networks. However, they are limited by the fact that only turn-on can be controlled (turn-off occurs spontaneously when the current cancels out). The slow nature of thyristors means that turn-off must be maintained for a constructor-specified minimum period T_q in order to avoid spontaneous restarting, as in the case of diodes. The symbol and characteristic $I_{Th}(V_{Th})$ for a thyristor are shown in Figure 1.11. We see that thyristors are three-segment switches, with unidirectional current (as in the case of a diode; the two share a symbol) and bidirectional voltage.

1.2.2.2. *The GTO thyristor*

The GTO thyristor is a development of the thyristor which aims to correct the main default of the basic component, i.e. the fact that it cannot be turned off by the controller. The GTO maintains the same current/voltage characteristic, but with bidirectional switching between the current and voltage segments (see Figure 1.12 – and the GTO symbol). Note that the GTO thyristor also shares the slow nature of classic thyristors, leading to the substitution of IGBTs wherever possible.

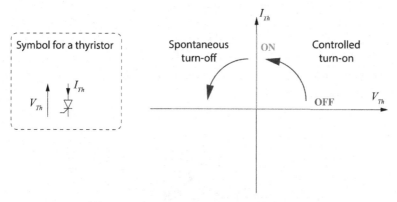

Figure 1.11. *Presentation of a thyristor*

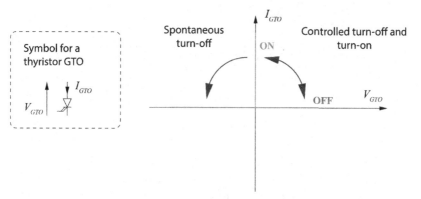

Figure 1.12. *Presentation of the GTO thyristor*

GTO thyristors are particularly difficult to control, as we need full mastery of the di/dt and dv/dt at the terminals during the switching phases, making a switching assistance circuit practically indispensable (see Chapter 3) for safe operations (i.e. operation in the SOA). This is a major drawback of the GTO compared to IGBTs or MOSFETs, for which the SOA is almost rectangular for rapid switching (as shown in Figure 1.10 for IGBTs in the previous section).

Furthermore, the trigger current needed for these components can reach hundreds of amperes (for a switched main current of several thousand amperes). For this reason, manufacturers currently offer components known as Integrated Gate-Commutated Thyristor (IGCTs) (see Figure 1.13), which combine a GTO thyristor with integrated controls to simplify the task of users designing converters for very high power levels.

Figure 1.13. *IGCT = GTO + integrated controls (source: ABB)*

1.2.2.3. *Antiparallel diode transistors*

The association of two components allows us to obtain a characteristic which is the combination of the two separate components. Thus, a transistor associated with an antiparallel diode, as shown in Figure 1.14, presents a characteristic which is reversible in terms of current and unidirectional for voltage. Note that the antiparallel diode is physically added to bipolar transistors and IGBTs, whereas an intrinsic structure diode exists in MOSFET transistors. This is clearly visible in Figure 1.8, where a PN junction may be observed between the drain and the source (more precisely the substrate). However, this diode is best avoided in practice, as it gives poor performance in terms of reverse recovery (as a "slow" diode with a significant stored charge).

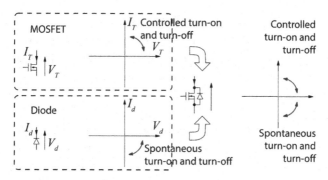

Note: the convention used for the diode is the opposite of the usual conventions

Figure 1.14. *Presentation of the antiparallel diode transistor*

1.2.3. *Four-segment switches*

1.2.3.1. *The triac*

The triac is the component which, at least in appearances, comes closest to the ideal four-segment switch, as it is bidirectional for both voltage and current (see Figure 1.15). However, triacs can only be controlled during turn-on, and turn-off occurs when the carried current cancels out, as in the case of thyristors. Triacs are, in fact, comparable to two antiparallel thyristors. They are used in the production of low-cost dimmers and for a wide range of domestic applications. The field of application for triacs is relatively limited (their slow speed means they are only suitable for "network" converters), but these components are important in semiconductor manufacturing as they are used in widely-distributed products (e.g. white goods).

1.2.3.2. *Synthesis using transistors and diodes*

It is possible to synthesize a four-segment switch by combining two transistors and two diodes, as shown in

Figure 1.16. This enables, for example, the production of Pulse Width Modulation (PWM) dimmers which are able to function at high switching frequency, unlike dimmers using triacs or thyristors. The increased complexity of these converters does limit their use, but this type of converter performs well, notably in terms of electromagnetic compatibility (EMC), as it is able to ensure absorption of a sinusoidal current in the network (on condition that a HF filter is used between the converter and the network). Synthesized four-segment switches will be covered in greater detail in Chapter 1, which focuses on AC/AC converters (more specifically in the context of matrix converters).

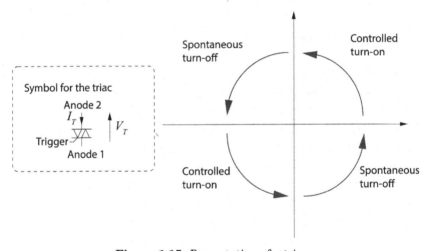

Figure 1.15. *Presentation of a triac*

1.3. Converter synthesis

1.3.1. *General approach*

In synthesizing a converter, we use the source interconnection rules defined above while focusing on the functions required for the target application. To do this, we must determine whether or not a source is reversible. For example, a *battery* may be recharged by inverting the direction of current through the cell: this is a *reversible current voltage source*. Similarly, electrical machines are generally made up of windings (induction circuits) which are assimilated to current sources; in the case of a *direct current machine*, we know that power can be supplied to the induction winding using a positive or negative voltage, conditioning the direction of rotation, and a positive or negative current, which determines whether the motor or generator (brake) mode is used. This is a *current source with reversible voltage and current*. Depending on the target application, these voltage reversals may or may not be necessary, and have a direct impact on the complexity of the converter.

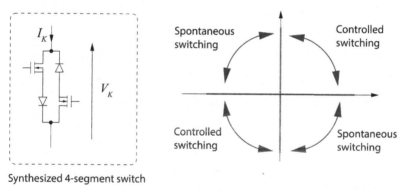

Synthesized 4-segment switch

Figure 1.16. *Four-segment converter using transistors and diodes*

1.3.2. *Application I*

As an example, we will consider the case of a direct current machine used for a hoisting application. Let us suppose that the power supply for the system comes from a battery. In this case, the machine is used in two operating quadrants, as two directions of rotation are required, but the machine couple is always oriented in the same direction. As the voltage is proportional to the speed (as a first approximation) and the couple is proportional to the current, we may state that the converter must be capable of voltage reversal, but does not need to be able to reverse the current at load level. However, it must be possible to recharge the battery during load descent phases (generator – i.e. brake – function of the motor), and therefore the current needs to be reversible at this level.

As we have two sources of different types, direct interconnection is possible, and we may create a list of authorized configurations (see Figure 1.17).

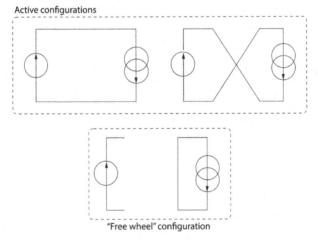

Figure 1.17. *Possible interconnection configurations for application I*

The H-bridge topology of the converter shown in Figure 1.18 allows us to obtain the three required configurations. Based on this observation, we need to define the four switches of the H-bridge as a function of the possible current directions, and establish the characteristics $I_K(V_K)$ of each switch. Once this task is completed, we must simply choose the most suitable switch options from the list given in Chapter 2.

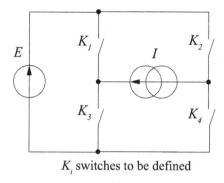

K_i switches to be defined

Figure 1.18. *H-bridge topology of the converter*

After a rapid examination characteristics required in the four switches, we obtain the result shown in Figure 1.19. This converter is known as a two-quadrant chopper, and will be discussed, alongside other possible configurations (one quadrant, two quadrants with reversible current, and four quadrants) in Chapter 1 of Volume 2.

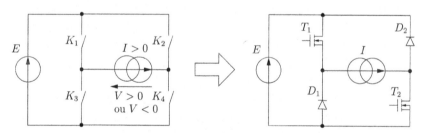

Figure 1.19. *Two-quadrant chopper with bidirectional voltage and unidirectional current*

1.3.3. *Application II*

Another classic application is for rectification (see Volume 2, Chapter 3 on AC/DC converters). In this case, we will consider a single-phase rectifier, which will be qualified later as a single-phase full-wave diode rectifier. Here, we will simply consider the topology required for this converter based on the nature of the interconnected sources:

– an alternating voltage source – AC (bidirectional voltage and current);

– a direct current source – DC (unidirectional current and voltage).

As in the previous example, we may make a list of authorized configurations; in doing this, we note that the results are the same as those for application I, so we may reuse the layout shown in Figure 1.18. Then, we need to consider the nature of the switches required in this case. The final result is the structure shown in Figure 1.20 (left), which can then be reorganized into a classic form (right) known as a *Graetz bridge rectifier*.

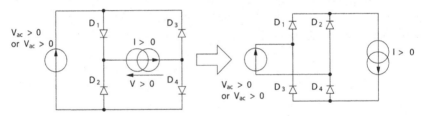

Figure 1.20. *Single-phase full-wave diode rectifier*

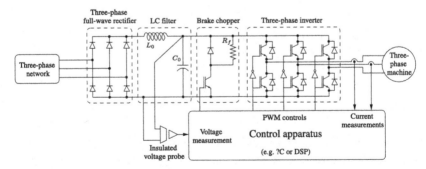

Figure 1.21. *"Classic" power layout of an industrial variable speed drive*

1.4. Analysis of a three-phase industrial variable speed drive

Let us return to the case of the industrial variable speed drive shown in the introduction (Figure I.1). The diagram is shown again in Figure 1.21. We clearly see how the interconnection of different elements satisfies the rules discussed above. The structure includes a diode rectifier (three phase, unlike that shown in Figure 1.20) which is connected to an AC power supply (ideal voltage source) and to a current source due to the presence of inductance L_0 in the LC filter on the DC bus. It also includes an inverter, connected to a three-phase machine (of the current source type) and, even without detailed knowledge of inverters, we see that an input voltage source is required: this is also present in the form of capacitor C_0. We also note the inclusion of a brake chopper, in parallel to the inverter, which is connected to a resistor and ensures dissipation of the energy sent back to the DC bus by the inverter (showing the presence of a reversible current) by the machine during phases of rapid deceleration. The chopper is needed due to the non-reversibility of current in the DC output of the rectifier, meaning that energy sent back to capacity C_0 will be trapped, eventually leading to an explosion if this energy is not consumed. Note the presence of a free wheel diode in the chopper. This diode is required for two reasons:

– the dissipation resistor is generally offset to avoid heating the variable speed drive, and to allow the use of fans for cooling where needed. As the connections are necessarily inductive, a free wheel is essential to avoid overcharging when the transistor is opened;

– the resistor often consists of a wire wound around a ceramic support with layers of wire wound in opposite directions in order to cancel out the magnetic field and the accompanying stray inductance as far as possible. However, some stray inductance is bound to remain, and this constitutes another reason for the use of a free wheel diode.

1.5. Study framework for converters

Different types of converters will be analyzed over the course of the following chapters. These systematically involve the connection of a voltage source and a current source. Depending on the specific case, the current source will either be presumed to be ideal, or obtained by the association of an inductor used in series with another dipole. Using this model, we will consider two operating types:

– *continuous conduction;*

– *discontinuous conduction.*

Continuous conduction occurs when the current in the inductor never cancels out, meaning that there is always a switch in ON mode. The voltage at the load terminals (particularly in the case of choppers) can therefore be determined (in the case of an ideal converter) using the conduction time intervals of the switch alone.

In the case of discontinuous conduction, the current in the inductor cancels out, and consequently, in a switching period, there will be a time interval during which no switch allows current to pass. In this case, the voltage at the load terminals no longer depends simply on the conduction time intervals of the switches, but also on the average current flowing through the load.

It is important to be aware of the study hypotheses used in the following converter studies:

– the time taken for switching is presumed to be null (instant switching – zero switching loss);

– switches are presumed to be ideal (no voltage drop in the ON state).

When analyzing converters, these hypotheses are reasonable as long as the switches are chosen with care (it would be unreasonable to use a 1,200 V IGBT with a voltage drop of over 2 V in the ON state for applications using a 12 V battery, for example). The impact of the voltage drop at the switch terminals must be negligible in order for the converter to function correctly. Losses in the switches are always analyzed *a posteriori*, using the current waveforms obtained when considering an ideal converter: this means that the models used can remain simple, with calculations which can be carried out analytically. If we were to take account of all of the characteristics of all of the components used, our task would be much more complicated and formal resolution would no longer be possible: in this case, computer simulation (using a program such as SPICE) would be necessary for studying converters.

2

Thermics, Packaging and Power Component Technologies

2.1. Losses in real components

2.1.1. *Notion of conduction losses*

While close to ideal switches in terms of switching modes, real components are subject to a voltage drop in the ON state, creating losses known as "conduction loss". This term does not include losses in the OFF state, as classic components are generally considered as ideal when open (OFF). There is no leaked current in circulation (or this current is negligible). In this section, we will, therefore, focus on losses in the ON state, specifically for two broad categories of components:

– bipolar components (bipolar transistors, insulated gate bipolar transistors (IGBT) and diodes);

– metal oxide semiconductor field-effect transistor (MOSFET) transistors.

A bipolar component in ON mode may be considered as a voltage source (ON state voltage V_{ON}, labeled V_{CEsat} for transistors and V_F for diodes in technical documentation) in series with a low dynamic resistance r_D (often omitted). Using this model, we can propose an expression of the average power P_{cond} dissipated by the component in a switching period T_d during which the component is in the ON state for a duration $\alpha.T_d$, where α is the *duty ratio*. For calculation purposes, we will suppose that, during the conduction phase, the current I flowing through the component is constant (in practice, it modulates, but this modulation is often negligible). We can, therefore, write:

$$P_{cond} = \frac{1}{T_d} \int_0^{\alpha.T_d} (V_{ON} + r_D.I).I.dt$$

$$= \alpha.\left(V_{ON}I + r_D.I^2\right) \qquad [2.1]$$

In the case of the MOSFET transistor, this value may be assimilated to a simple resistance in the ON state, written R_{DSon} in manufacturer documentation. Consequently, the power dissipated by conduction is written simply:

$$P_{cond} = \alpha.R_{DSon}.I^2 \qquad [2.2]$$

2.1.2. *Notion of switching loss*

Switching losses are not only dependent on the switching component, but also on the environment. For illustrative purposes, we may consider the example shown in I.2, corresponding to the characteristic in Figure I.3. We see that, during switching, the point of operation moves from point S (saturated transistor) to point B (transistor turned off) or vice versa, but this move is not instantaneous, and, during switching, the characteristic of the transistor takes forms that vary continuously between those referenced by the commands V_{BE1} and V_{BE6}. Thus, when the command is V_{BE5}, the instantaneous point of operation of the construction is M, where we observe a current I_{T0} and a voltage V_{T0} coexisting at the transistor terminals. The instantaneous power in the transistor is, therefore, null, and losses are experienced, in addition to those due to conduction.

As a first approximation for study purposes, we may assume that point S lies on the ordinate axis (no voltage drop at the transistor terminals in the ON state). If we state that, during switching (of duration T_c), the point of operation M of the transistor varies in a linear manner as a function of time in the segment with $[BS]$, we obtain:

– for turn-on (from B toward S):

$$\begin{cases} i_T(t) = \frac{E.t}{R.T_c} \\ V_T(t) = E\left(1 - \frac{t}{T_c}\right) \\ P_T(t) = V_T(t).I_T(t) = \frac{E^2.t}{R.T_c}\left(1 - \frac{t}{T_c}\right) \end{cases} \qquad [2.3]$$

– for turn-off (from S toward B):

$$\begin{cases} i_T(t) = \frac{E}{R}\left(1 - \frac{t}{T_c}\right) \\ V_T(t) = \frac{E.t}{T_c} \\ P_T(t) = V_T(t).I_T(t) = \frac{E^2.t}{R.T_c}\left(1 - \frac{t}{T_c}\right) \end{cases} \qquad [2.4]$$

Consequently, the average switching losses P_{comm} (for period T_d) are expressed as

$$P_{comm} = \frac{2E^2}{R.T_d} \int_0^{T_c} \frac{t}{T_c}\left(1 - \frac{t}{T_c}\right).dt \qquad [2.5]$$

Using $\tau = t/T_c$ (and thus $dt = T_c.d\tau$), we obtain

$$P_{comm} = \frac{2E^2.T_c}{R.T_d} \int_0^1 \tau(1 - \tau).d\tau = \frac{4E^2.T_c}{3R.T_d} \qquad [2.6]$$

Switching needs to occur as quickly as possible to reduce switching losses.

Unfortunately, in practice, powered loads (particularly electrical machines) are inductive and require the use of a free wheel diode[1] (see Figure 2.1) and in this case the possible reduction in switching time T_c is limited: the diode is subject to the reverse recovery phenomenon discussed in section 1.2.1.1. This point is critical, except in the case of Schottky diodes, as rapid switching increases the downward slope $\frac{dI_D}{dt}$ of the current in the diode while the transistor turns on. As the stored load is the integral of the current in the diode, for a fixed evacuated charge, the inverse current will reach an amplitude that increases with the slope $\frac{dI_D}{dt}$. This current peak in the diode has repercussions in the transistor when the load is an ideal current source. Consequently, it must be taken into account when choosing the rated current of a transistor, as excessive switching speeds could lead to the destruction of the component.

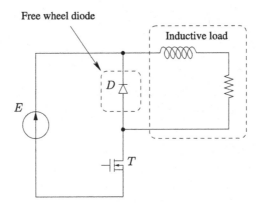

Figure 2.1. *Switch and inductive load – free wheel diode*

2.2. Semiconductors in power electronics

Silicon is the most widely used material for synthesizing components for power electronics, but other technologies also

---

1 A free wheel diode is needed to ensure the continuity of current in the load, as otherwise overloading would occur at the load terminals and might damage the transistor.

exist at a variety of levels of maturity. These other materials are known as *wide-bandgap semiconductors*, characterized by the energy difference (denoted as eV) separating the *valence band* and the *conduction band* (known as the *bandgap*), wider than that encountered using silicon.

	Classic semiconductor		Wide-bandgap semiconductor				
	Si	GaAs	3C SiC	6H SiC	4H SiC	GaN	C
Bandgap (eV)	1.12	1.4	2.3	2.9	3.2	3.39	5.6
Mobility μ_n of e^- (cm^2.V^{-1}.s^{-1})	1,450	8,500	1,000	415	950	2,000	4,000
Mobil. μ_p of holes (cm^2.V^{-1}.s^{-1})	450	400	45	90	115	350	3,800
Critical electric field E_c (V.cm^{-1})	3×10^5	4×10^5	2×10^6	2.5×10^6	3×10^6	5×10^6	10^7
Thermal cond. λ (W.cm^{-1}.K^{-1})	1.3	0.54	5	5	5	1.3	20
Max. temp. for use T_{max} (°C)	125	150	500	500	500	650	700

Table 2.1. *Principal characteristics of semiconductor materials*

Table 2.1 compares the different classic (Si and GaAs) and wide-bandgap materials (SiC – several variants, GaN, C – i.e. diamond) in terms of the bandgap, carrier mobility (electrons and holes), the critical electrical field, thermal conductivity and the maximum temperature for use. For the purposes of power electronics, the following characteristics of semiconductors are of interest:

– thermal conductivity;

– heat withstand;

– voltage withstand.

Significant gains can be made on these points with minor sacrifices in terms of electron mobility (note, however, that hole mobility is considerably reduced). These materials, therefore, offer a variety of benefits:

– low losses;

– losses are evacuated well due to the strong thermal properties of the materials;

– bulky heat sinks are not required as the maximum acceptable temperatures at junction point are high.

These performances at "component" level produce interesting results at "system" (and therefore application) level:

– overall miniaturization of converters (notably of the cooling element);

– the possibility of operating in a thermally difficult environment ("high temperature" electronics).

Figure 2.2 shows an extract from a data sheet for a SiC MOSFET transistor, developed by Micross Components, with a maximum acceptable junction temperature of 210°C.

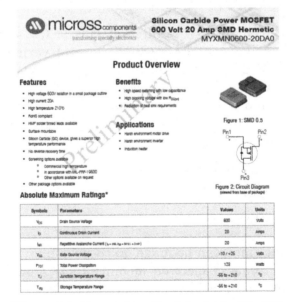

Figure 2.2. *Extract from the data sheet of the SiC MYXMN0600-20DA0 MOSFET transistor (source: Micross Components)*

Diodes are the "reference" component in terms of performance targets for these technologies, as they are the simplest components and have been produced in SiC form on an industrial scale for some time now (the production of more complex components, such as MOSFETs and Junction Field Effect Transistor (JFETs), in this way is a more recent development). Figure 2.3 shows the current waveforms during the turn-off phase for different diodes, including an SiC Schottky diode (the others are silicon diodes with classic, "ultra-fast" type PN junctions). The Schottky diode is the only example not to present an inverse current. As we have seen, this characteristic is interesting for the diode itself (no switching losses during reverse recovery) and for the environment (no current peaks in addition to the current in the load in the chopper shown in Figure 2.1). This type of diode also exists using Si technology, but is limited to reverse voltages of the order of 200 V; high-rated voltages can easily be used using SiC technology, which has a higher critical electrical field level.

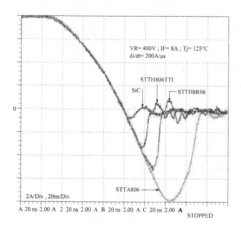

Figure 2.3. *Comparison of the turn-off phases of different diodes (source: ST Microelectronics – R. Pezzani). For a color version of the figure, see www.iste.co.uk / patin / power1.zip*

2.3. Packaging of power electronics components

2.3.1. *Discrete components*

Discrete components contain a single component (or two components in the case of transistors with antiparallel diodes) in a simple package, usually designed to be soldered to a printed circuit. Two forms of discrete component packages are generally used (see Figure 2.4):

– "through-hole" designs (type TO220), which must be fixed to a printed circuit through holes;

– "surface-mounted" devices (SMDs), which are fixed onto a printed circuit, using glue where necessary; the pins are soldered onto the copper pads on the surface of the circuit.

Discrete components (MOSFET, IGBT, diodes)
and packages: TO220, SMD, etc.

Figure 2.4. *Through-hole and SMD packages for discrete components (source: International Rectifier)*

The latter technique is becoming increasingly widespread, as it makes efficient heat dissipation possible via the printed circuit (ground plane) and improves the productivity of circuit board assembly lines (SMD components are quicker to place than through-hole designs using automated methods), thus reducing production costs. They are also used in specialist circuits using a metallic substrate. In this case, the printed copper circuit is insulated from the metal base using a layer of insulating material, which is either organic or obtained using a metal oxide (alumina). This technique is similar to that used in the power modules presented in the next section.

2.3.2. *Power modules*

Power modules are designed to be fixed to a heat sink, and contain several integrated components. They are better suited for medium- and high-power applications than discrete components, and greatly simplify the wiring of the "power" part of a converter, which is partially or completely prewired (some modules can contain all the power components needed to produce an industrial variable-speed drive, such as the one shown in Figure I.1, with the exception of the passive components, i.e. coils and capacitors).

The module has a metallic (aluminum) base to enable it to be fixed to the heat sink. This is covered with a dielectric (e.g. a ceramic element) and with copper leaf, onto which the dies (diodes, transistors, etc.) are soldered. These chips are then interconnected using fine aluminum wires[2] (known as "bonding" wires). This technology is known as *Direct Bond Copper* (DBC), as the chips are connected to a copper surface in the same way as a classic printed circuit (this technology is also offered by the manufacturers of traditional printed circuits, such as Rogers Corp. – offering the *Curamik range*). The front surface of the module carries all the electrical connections (for both power and control purposes) required by the environment. At medium and high powers, connections are established using screw terminals to affix an aluminum

---

2 Gold bonding wires are only used in cases where very fine wires are required (gold has a high ductility level), for example, in signal processing chips, but not for power electronics.

or copper bus bar carrying filtering capacitors (see Figure 2.5). In certain cases, some of the connections are made using spring contacts, designed to be attached to printed circuits (see Figure 2.6): this technique is particularly suitable for onboard applications, as the spring contacts withstand vibrations better than traditional soldering.

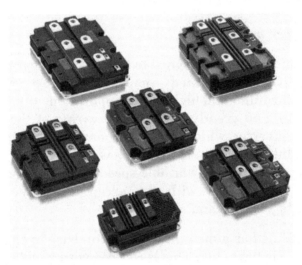

Figure 2.5. *(High) power modules (source: ABB)*

Figure 2.6. *Power module with spring contacts (source: Semikron)*

2.4. Thermal modeling of components

2.4.1. *General points*

As we have seen in section 2.1, semiconductor components are subject to losses through conduction and in switching. These losses lead to heating which, if left unmanaged, can destroy the component. We, therefore, need to analyze this problem to enable correct dimensioning of the associated cooling equipment; this is carried out using thermal modeling of components. Generally speaking, heat propagation involves three distinct physical phenomena:

– *conduction*, where atoms transmit vibration to each other due to heat[3];

– convection, where currents are created in a fluid surrounding a heated element (in this case, current refers to the movement of molecules, not an electrical current), linked to density changes in relation to temperature;

– radiation, which, unlike the first two phenomena, does not require a physical support, as the heated element emits an electromagnetic wave (carrying a certain energy). This phenomenon has been particularly well modeled for the "theoretical" case of the black body, where the density of the emitted power can be expressed as a function of the temperature T (following a law of the form T^4, known as the *Stefan–Boltzmann law*).

In this case, we will use a purely linear thermal modeling, which corresponds closely to the hypothesis used by manufacturers, i.e. the non-inclusion of the radiation

---

3 Temperature, a macroscopic value, is the translation to "human" level of the vibration of atoms and molecules making up a material around their balance position. This erratic movement of small scale (micro- or mesoscopic) particles is known as *Brownian motion* (named for the Scottish botanist who observed the phenomenon for the first time in 1827 using pollen suspended in a liquid).

phenomenon (which is fundamentally nonlinear, as shown by the Stefan–Boltzmann law). In this context, the domain of thermics is analogous to the electrical domain, insofar as we are able to establish a connection between thermal values (heat flow and temperature) and electrical values (current and voltage). It is, therefore, logical that we should use relationships similar to Ohm's law connecting the heat flow in watts (equivalent to a current) with a difference in temperature expressed in K or °C (equivalent to a voltage) via a thermal resistance (defined in °C/W).

2.4.2. Steady-state thermal analysis

For the purposes of steady-state thermal analysis, we will not consider temperature changes over time in the course of an operating cycle of the system in question. We simply wish to determine the temperature(s) of the system at the nominal operating point. Our model, therefore, comes down to a more or less complicated network of thermal resistances (characterizing the thermal interfaces between system elements) and thermal power sources (comparable to current sources in electrical terms) symbolizing the losses in components. In an equivalent electrical circuit, note that each node i is at a certain voltage T_i defined *in relation to a reference value*. In an electrical circuit, we generally define a *plane* which, by convention, is a voltage of 0 V; in the case of a thermal study, the reference is a *thermostat*, the temperature of which is presumed to be independent of the heat flow it absorbs. This may be a solid metal support, but *in many cases*, the *surrounding air* is considered to be the thermostat; this hypothesis is, however, *not always ideal* (particularly for equipment in a small, confined space).

To illustrate the modeling of a real system, we will use the minimal structure shown in Figure I.2, i.e. a transistor in series with a purely resistive load powered by a voltage source E. In the case of a bipolar transistor (leaving aside losses in the Base-Emitter junction), the expression of the total losses P_T in the transistor has already been established:

$$P_T = P_{cond} + P_{cond} \simeq \alpha. \left(\frac{V_{ON}.E}{R} + \frac{r_D.E^2}{R} \right) + \frac{4E^2.T_c}{3R.T_d} \qquad [2.7]$$

The parameters E, R and α are determined by the application, while the designer selects a transistor (and, consequently, V_{ON} and r_D) and the times T_d and T_c (with consideration for the constraints imposed by the transistor and the application).

Once a transistor has been chosen (and the losses have been evaluated), thermal parameters may be obtained from the manufacturer's documentation, including:

– maximum junction temperature θ_{jmax};

– the thermal resistance between the junction and the surrounding air R_{thj-a};

– the thermal resistance between the junction and the package R_{thj-c}.

The thermal resistance between the junction and the surrounding air corresponds to the capacity of the package to evacuate heat. As this capacity is limited, the value is significantly higher than that between the junction and the package. We must also take into account the contact resistance between the package and the heat sink (R_{thc-s}) which depends on:

– correct attachment of the heat sink (mounting torque loads specified for power modules);

– the surface state of the two elements in contact (use of a reasonable quantity of thermal grease is highly recommended to maximize heat transfer (see Figure 2.7);

– the use, or otherwise, of a sheet of electrical insulation between the component and the heat sink (e.g. a sheet of mica (see Figure 2.8) which poses a partial obstacle for heat transfer.

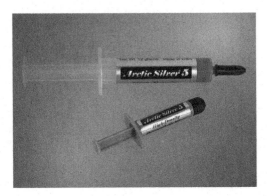

Figure 2.7. *Thermal grease syringes (source: Artic Silver)*

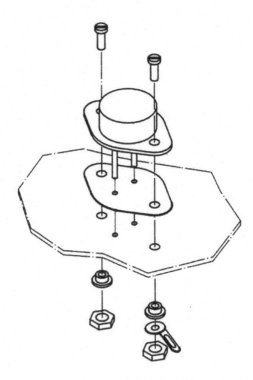

Figure 2.8. *Electrical insulation kit, including a leaf of mica for a TO3 package (source: Fischer Elektronik)*

The heat sink itself is then characterized by a thermal resistance in relation to the surrounding air, denoted R_{ths-a}.

We thus obtain the equivalent diagram shown in Figure 2.9, where we know all of the thermal resistances except for that of the dissipator R_{ths-a}, the power P_T, and the ambient θ_a and nominal junction temperatures θ_j which should preferably be below θ_{jmax} to guarantee the reliability of the converter[4]. Note, however, the absence of the thermal resistance R_{thj-a} between the junction and the surrounding air at component package level, as the junction is directly in parallel with the power source pT and therefore with the association $R_{thj-c} + R_{thc-s} + R_{ths-a}$. We know that this value is high, unlike the other thermal resistances in the circuit; this means that the value is negligible. These considerations allow us to calculate the thermal resistance required for the heat sink:

$$R_{ths-a} = \frac{\theta_j - \theta_a}{P_T} - R_{thj-c} - R_{thc-s} \qquad [2.8]$$

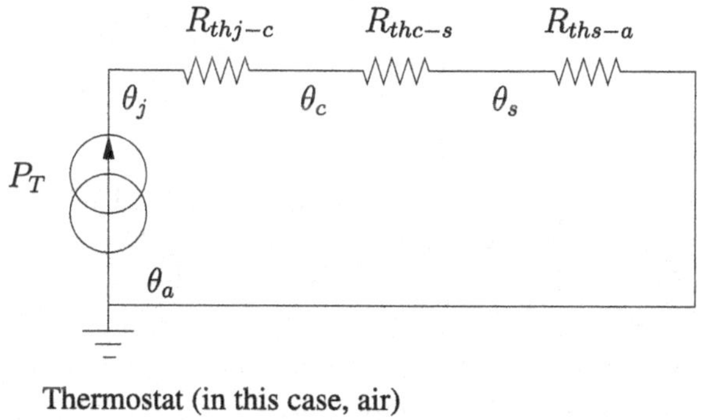

Thermostat (in this case, air)

Figure 2.9. *Electrical diagram equivalent to the thermal model of the transistor in established phase*

---

4 The lifespan of a transistor is linked to the junction temperature, and more precisely to temperature variations over time, introducing thermo-mechanical constraints which can lead to fissures at various points in the component (metallic base, bonding wires, etc.).

In equation [2.8], we see that the value obtained can be negative if the dissipated power P_T is too high. In this case, cooling is impossible, unless we use a thermostat cooler than the ambient temperature (e.g. a Peltier effect module).

2.4.3. Transient mode

In certain applications, the duration of usage T_u of the converter is low in comparison with the idle time T_i. In this case, the cooling system may be designed, not for the permanent cooling of component losses, but to prevent the temperature of the component(s) from exceeding a certain temperature after time T_u. The model presented in the previous section is no longer sufficient; we need to take account of the *dynamics of temperature evolution* in the equivalent circuit. To do this, we use the notion of *thermal capacity* (in J/K or J/°C). In theory, all materials have a specific heat capacity; for example, water has a specific heat capacity of 4.28 kJ/kg/°C. If we know the mass of all the elements in the system and the materials used, we should easily be able to evaluate the thermal capacities involved in the electrical circuit equivalent to the thermal model. However, in practice, only the heat sink (the largest, and often heaviest, component) needs to be modeled in this way, as its capacity is normally (by far) the highest. As an indication, note that the specific heat capacity of aluminum (the standard material for heat sinks) is 904 J/kg/°C. When a heat sink is selected (aluminum profile), charts are used to obtain the thermal resistance for a length l of a given profile. We can also obtain the mass per unit length. In this way, we obtain a two-component thermal model of a heat sink (resistance + thermal capacity), which are functions of a single parameter: length l. We must then consider the dynamic model in Figure 2.10 to see whether, for a given usage profile (use over a period of time T_u, idle time T_i), the junction temperature of the transistor remains below the predetermined value.

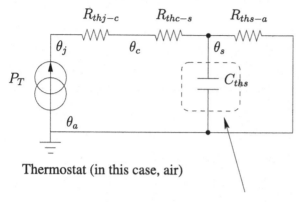

Thermostat (in this case, air)

Thermal capacity of the heat sink

Figure 2.10. *Electrical diagram equivalent to the dynamic thermal model of the transistor*

2.5. Choosing a heat sink

Heat sinks are selected primarily on the basis of the required thermal resistance, but the type of package used for the cooled component may also be taken into consideration. Different heat sink designs are better suitable for the specific form of certain components (e.g. the heat sink shown in Figure 2.11).

Figure 2.11. *A heat sink and the corresponding component package (TO-3)*

Modules with metallic bases are becoming increasingly widespread in (high) power electronics. This simplifies the choice of heat sinks, as aluminum profiles may be used systematically in this case; designers must simply determine the length required for a given thermal resistance (see Figure 2.12).

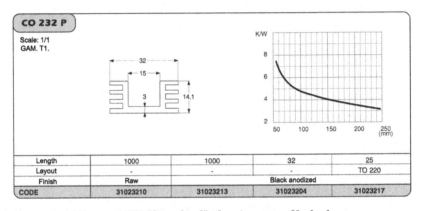

CO 232 P			
Length	1000	1000	32	25
Layout	-	-		TO 220
Finish	Raw		Black anodized	
CODE	31023210	31023213	31023204	31023217

Figure 2.12. *"Standard" aluminum profile for heat dissipation (taken from the Seem catalog)*

Thermal resistance values should be taken as *qualitative indications* rather than *precise quantitative data,* as the orientation of the fins has a significant effect on the results. A heat sink will attain maximum efficiency when the (vertical) movements of air convection are unobstructed; for this reason, fins are best oriented vertically, as otherwise the thermal resistance may increase by 20%. The results obtained will also vary depending on whether the heat sink is painted black or left plain (in this case, the thermal resistance may be increased by 10%[5]). Finally, it would be foolish to think that a heat sink of 1 m in length would be

---

5 This value, given by certain manufacturers, should be kept in perspective, as the impact of the paint only effects dissipation by radiation, which is fundamentally nonlinear (expressed in T^4) and cannot be modeled, except around an operating point, by an equivalent thermal resistance.

efficient for the dissipation of power generated by a component of 1 cm in length: the localized constant model given here cannot be applied to non-compact equipment. In these cases, only finite element simulations will be able to produce satisfactory results[6].

A final point to address is the mechanism for heat exchange between the heat sink and the air. Previously, we have considered exchanges by natural convection (hence, the need to correctly align the fins in order to obtain maximum efficiency). Forced air circulation is also possible with the use of fans. In this case, we may use:

– charts for correction (downward) of the required profile length;

– thermal resistance tables for a given heat sink according to the type of air circulation (natural or forced).

In the cases where forced air circulation is used, the speed of circulation (in m/s) is a key factor, as it enables us to evaluate the obtained thermal resistance.

2.6. Other types of cooling

2.6.1. *Liquid and two-phase cooling*

To obtain compact and efficient cooling solutions, air may be replaced by a cooling fluid of higher density, such as water, in order to improve the heat exchange. This type of cooling is particularly widespread for onboard applications, notably in the automobile industry. This type of cooling solution is also used for computer components for reasons of space, efficiency and noise reduction when compared to fan-based methods (see Figure 2.13).

---

6 On the condition that the system parameters are perfectly known, something which is often difficult in a thermal study.

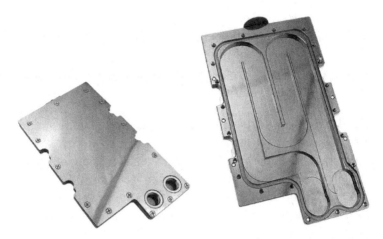

Figure 2.13. *Cooling plate for 3.5 in. hard drive*
(taken from the Koolance catalog)

Another particularly efficient cooling technique uses the notion of latent heat during phase changes. As an illustration, let us consider the example of melting ice: when ice melts at 0°C, heat energy is absorbed with no variation in temperature. The temperature remains at 0°C for as long as the ice is not completely melted: this is a latent fusion phenomenon. The same phenomenon occurs when water evaporates, and, more generally, for all materials during a change of phase (solid $\Longleftrightarrow$ liquid or liquid $\Longleftrightarrow$ gas). The liquid $\Longleftrightarrow$ gas transition is particularly interesting, as we can use natural circulation (via a heat-pipe – see Figure 2.14) of gas toward a condenser in which the gas evacuates stored heat and returns to its liquid state before returning to the exchanger connected to the heat source.

Water may be used in this type of equipment; in this case, the evaporation temperature can be reduced by working at lower pressure (e.g. 0.05 bar to obtain a phase change temperature of 33°C). However, ammonia is more widely used, with a vaporization temperature of 15°C under normal atmospheric pressure (this temperature is higher under higher pressure) and a latent vaporization heat of 1,371.2 kJ/K at the same pressure (in comparison, the latent vaporization heat of water is around 2,600 kJ/K for a wide range of pressures). In certain specific applications (sometimes dimensioned for purely transitional operations), paraffin may be used in cooling systems (Eicosane, with a fusion temperature of 36.6°C and latent fusion temperature of 243 kJ/K).

Figure 2.14. *Heat pipe for microprocessors*
(Cooler Master TX3)

2.6.2. *Active cooling*

Another type of "active" cooling involves the use of junctions between conductors of different natures, through which a current is passed. This current triggers heat transfer due to the Peltier effect. Peltier effect modules are available for use in cooling electronic components, as illustrated in Figure 2.15 (in this case, in addition to liquid cooling).

Note, however, that this type of apparatus is generally used for reducing the temperature of analog electrical circuits in order to reduce noise in metrological applications where precision is essential. These techniques may still be used in power electronics, and the technology offers certain advantages:

– the ability to reduce the temperature of the cold plate to below the ambient temperature;

– totally static operation (no circulation of fluid), which is not sensitive to accelerations or vibrations (advantageous in aeronautic or spatial applications, notably in the military context).

Figure 2.15. *Mixed liquid / Peltier effect cooling (Swiftec)*

3

Auxiliary Converter Circuits

3.1. Gate control in MOSFET and IGBT transistors

3.1.1. *Principles*

A metal–oxide–semiconductor field-effect transistor (MOSFET) (or insulated-gate bipolar transistor (IGBT)) is controlled by the application of a voltage between its gate and its source (or emitter): for voltages between zero and a few volts, the transistor is turned off, then once a threshold voltage (V_{GSth} or V_{GEth}) is reached, the transistor allows current to pass, at first in a linear mode, and then tending toward a state of saturation. To ensure this happens, a voltage significantly higher than V_{GSth} (or V_{GEth}) must be applied in order to guarantee the saturation: in practice, a voltage between 10 and 15 V is sufficient for this purpose. As the gate is physically equivalent to a capacitor frame, the transistor behaves in the same way from a control perspective. Consequently, during switching, the injected current may need to be quite high (up to several amperes) to ensure rapid switching, thus guaranteeing optimum switching losses. A driver circuit therefore needs to include an output stage capable of delivering these currents. Outside of switching phases, the gate current is null, MOSFET and IGBT control is said to be non-dissipative and the average power required to power the driver is relatively low (although instantaneous power may be high).

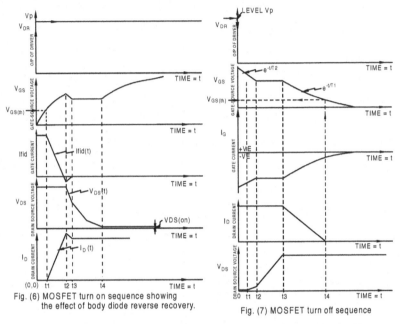

Fig. (6) MOSFET turn on sequence showing the effect of body diode reverse recovery.

Fig. (7) MOSFET turn off sequence

Figure 3.1. *Turn-on (left) and turn-off (right) in a MOSFET (source: IXYS documentation)*

As we saw in Chapter 1, switching needs to occur as quickly as possible in order to limit switching losses in transistors. However, we also need to limit the current variation slope during turn-on in order to reduce the inverse current peak in the opposite free wheel diode, which will turn off, as far as possible. The free wheel diode stores charges in a space-charge zone, so the negative current peak caused by turn-off increases as its duration decreases. The technical documentation of the selected free wheel diodes should be consulted when selecting a switching time, which should be approximately three to four times the increase time of the current in the transistor drain ($t_2 - t_1$ in Figure 3.1). Particular attention should be given to parameters I_{RM} (reverse recovery current) and Q_{rr} (reverse recovery charge) with the test conditions specified in the documentation. The most suitable technologies for these purposes are classic rapid, or ultra-rapid, recovery diodes, or, where possible (for low-voltage applications), Schottky diodes, for which the recovery phenomenon does not exist. Diodes dedicated to low-frequency rectifiers (connected to the grid – 50/60 Hz) should never be used in these situations, as they are not suited to the required switching speeds and the stored charge involved is too much high.

Once appropriate switching times (T_{on} for turn-on and T_{off} for turn-off) have been selected in relation to the constraints imposed by the diodes, we must simply choose the gate resistance(s) needed to carry out the switching operations required by the transistor.

Switching times can be approximated using a variety of methods, but these all (normally) produce very similar results. In this case, the "gate–source/emitter" dipole of the transistor will be approximated as a constant capacity throughout the switching phase. In a similar way, this capacity may be determined by noting the charge Q_g stored in the transistor gate when the voltage V_{GS} (or V_{GE}) reaches the final value V_{fin} imposed by the driver circuit. To do this, we use the characteristic $V_{GS} = f(Q_g)$ supplied in the manufacturer documentation. The graph shown in Figure 3.2 is taken from documentation of this type.

From the graph, we see that for a gate control V_{GS} of 10 V and with a voltage V_{DS} of 50 V, the charge stored in the gate reaches 200 nC. In these conditions, we can evaluate the equivalent gate capacity C_{geq} as 20 nF ($Q_g = C_{geq}.V_{GS}$). We may then simply consider that the switching time is of the order of time constant $\tau = R_g.C_g$ (in practice, between τ and 3τ), and thus deduce the gate resistor R_g to insert between the driver and the transistor, as shown in Figure 3.3.

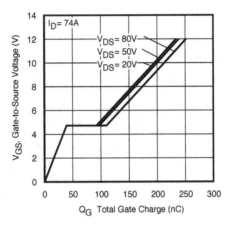

Figure 3.2. $V_{GS} = f(Q_g)$ *characteristic of an IRF7769L2 transistor (source: International Rectifier Documentation)*

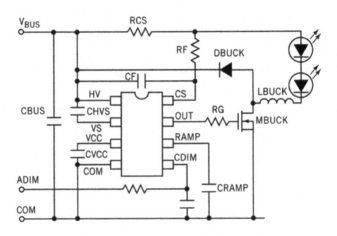

Hysteric Buck LED driver with low-side MOSFET

Figure 3.3. *Driver from a Buck chopper designed to power an LED – International Rectifier Circuit IRS2980*

Note that it is also possible to use two gate resistors for differentiated control of turn-on and turn-off in the transistor. To do this, we must simply place a diode in series in each resistor to enable current to circulate in one direction only (one resistor for gate charging, therefore for turn-on, and a second for discharging, i.e. turn-off). This function may be integrated into industrial drivers, such as that shown in Figure 3.4, used for controlling high-voltage IGBT transistors in a half-bridge and providing galvanic insulation between the control and power interfaces, surveillance of transistor terminal voltages in case of short circuits (desaturation of IGBTs), or to establish a deadtime in the control of two switches.

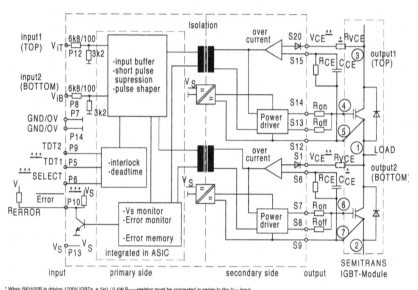

* When SKHI22B is driving 1700V IGBTs, a 1kΩ / 0,4W R$_{VCE}$-resistor must be connected in series to the V$_{CE}$ input.
** The VCE-terminal is to be connected to the IGBT collector C. If the V$_{CE}$-monitoring is not used, connect S1 to S9 or S20 to S12 respectively.
*** Terminals P5 and P6 are not existing for SKHI22A/21A; internal pull-up resistor exists in SKHI22A/21A only.
1-7 Connections to SEMITRANS GB-module

Figure 3.4. *IGBT half-bridge driver – Semikron SKHI21 or 22A / B circuit*

Taking a switching time of 400 ns, a gate resistance must be chosen from within the following interval:

$$3.33\,\Omega \le R_g \le 10\,\Omega \qquad [3.1]$$

This approach is not sufficient for a precise choice of switching time. We therefore need to use a Simulation Program with Integrated Circuit Emphasis (SPICE)-type simulation tool to refine our calculations, something that requires minimal time but presents high levels of precision. Note that the calculated gate resistance is not limited to the resistor placed between the driver and the transistor alone, as the transistor itself presents a resistance of access to the gate that is not negligible in cases using low gate resistance (rapid switching times). This phenomenon is particularly visible in "TrenchMOS" (trench MOSFET)-type transistors, which are the preferred component type for low-voltage applications. The internal gate resistance in the transistor can even prove problematic if we wish to dimension a low-power converter with a high switching frequency (>1 MHz) while retaining high global efficiency from the converter (gate drive losses can represent a non-negligible part of total losses in such cases).

3.1.2. *Controlling a "high side" transistor*

In the previous paragraph, we discussed gate control in simple cases of transistor control where the source (or emitter) is connected to the ground. Figure 3.4 shows a diagram of an IGBT half-bridge driver used to control a high side transistor; based on the studies already completed, we know that the emitter voltage can vary between the voltage of the ground and that of the positive terminal of the DC/not continuous bus (e.g. 600 V). In these conditions, a classic control circuit cannot operate properly: a "floating" power supply is needed to allow the circuit to maintain the voltage of the transistor gate a few volts (in practice, 10–15 V) above the emitter voltage (in this case, 600 V). Clearly, it is not practical to manufacture a 615 V control power supply; it is better to use an insulated 15 V supply, with a negative terminal connected to the floating voltage (that of the mid-point of the half-bridge). This objective may be fulfilled using two techniques:

– creation of a galvanically insulated switch-mode power supply with a 15 V output (flyback or forward);

– use of a bootstrap circuit.

The first solution appears simple, but requires a relatively complex and costly circuit, making this method unattainable for a considerable number of applications (however, it is used in the SKHI22 driver, for example). The second, more "rustic" solution, which is presented in Figure 3.5, is more widely used.

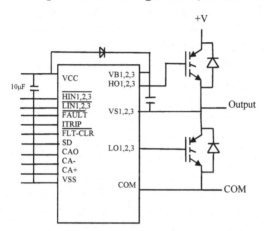

Figure 3.5. *Driver for a three-phase SHD830301 inverter (source: Sensitron Semiconductor)*

Obviously using this type of driver, control of the low side transistor is not an issue as its source (emitter) is connected to the ground. Moreover, the driver is able to supply the current required by the transistor gate in order to control the switching time (turn-on and turn-off). Note that although no gate resistance is shown in Figure 3.5, this is generally required (the figure is a simplified illustration). The key element for the control of the high side transistor in the circuit is the diode placed between terminals Vcc and VB1,2,3 (these are, in fact, three terminals corresponding to the drivers of the top three transistors of the three half-bridges of the inverter) and the capacitor (in reality, three capacitors) placed between terminals VB1,2,3 and VS1,2,3 (which are themselves connected to the three mid-points of the inverter half-bridges). The following reasoning applies to a half-bridge (and can be applied in the same way to the two others): when a command is given to the lower transistor, the converter output voltage is brought to the same level as the ground voltage (com). In these conditions, a current travels through the power assembly Vcc + diode + capacitor (presumed to be uncharged at the starting point), leading the capacitor to be charged with voltage Vcc. We say that the bootstrap component is triggered. If we then open the low side transistor and give the order to close the high side transistor, the output terminal voltage increases to tend toward $+V$. In practice, we obtain $+V \gg V_{cc}$, or at least $+V = V_{cc}$: in these conditions, it would not be possible to maintain satisfactory control of the high side transistor, as for the high side transistor to remain in ON state would require a gate voltage of $+V + V_{GSon}$ (with $V_{GSon} = 10$ or 15 V, for example). To achieve this result, the capacitor uses a floating voltage source Vcc to maintain the voltage of VB1,2,3 at a level equal to that of VS1,2,3 $+Vcc$. The driver then simply connects the charged capacitor between the gate and emitter of the transistor, via a galvanically insulated command inside the circuit.

It means that we do not only need to dimension the driver:

– as in the initial situation, it must be able to deliver the gate current needed in order to produce the required switching times;

– it must possess galvanic insulation that is compatible with the value $+V$ between the control circuit and the transistors;

– the external bootstrap diode itself must be chosen with reference to the voltage $+V$ (more precisely, it must be able to withstand a reverse voltage equal to $-V + Vcc$)[1];

---

1 It must also be able to withstand current peaks corresponding to the recharging of the capacitor for each switch period, and more precisely to each triggering of the lower transistor (or its antiparallel diode).

– the bootstrap capacitor must possess a capacity C_{boot} sufficient for correct control of the transistor.

This last point is crucial, in which C_{boot} stores a limited quantity of energy, and part of this energy will be transferred to the transistor gate: in these conditions, while the capacitor is initially charged under a voltage Vcc, the voltage at the terminals will drop when the charge transfer occurs. The transferred energy value must therefore be low in relation to the energy initially stored in the capacitor. The stored energy W_{elec} in a capacity C at voltage V is written:

$$W_{elec} = \frac{1}{2}CV^2 \qquad [3.2]$$

so we must ensure that:

$$\frac{1}{2}C_{boot}.Vcc^2 \gg \frac{1}{2}C_g.Vcc^2 \qquad [3.3]$$

i.e. put simply:

$$C_{boot} \gg C_{geq} \qquad [3.4]$$

In our previous example (where $C_{geq} = 20$ nF), a bootstrap capacitor of $2\mu F$ would be sufficient, for example (in practice, the value is $2.2\mu F$).

N.B.: On power-up, the bootstrap is discharged, and needs to be triggered by forcing the closure of the low side transistor of the half-bridge. If we attempt to start the control for the high side transistor, notably by hysteresis, the circuit will not work. Furthermore, a bootstrap requires a periodic recharge of capacity C_{boot}, as it is not really a voltage source, and energy loss occurs through the control process. Capacitors, in general, are subject to current leakage (particularly in the case of electrolytic capacitors). Consequently, a saturated control cycling at 100% over a long period of time may lead to system breakdown. These limitations, however, are balanced out by the advantages of bootstrap techniques, which allow simple and low-cost control of MOSFET and IGBT half-bridges (or, at the least, a high side transistor in a converter, i.e. a transistor using a source or emitter with floating voltage).

3.2. Snubbers

3.2.1. *Switching loss in converters*

Switching losses occur due to the simultaneous coexistence of significant voltage and current at switch terminals during the switching phase (turn-on or turn-off). Losses occur because the switch (generally a transistor) behaves in a linear manner during this phase; these losses may represent a non-negligible part of the total losses in the switch (of a similar order to conduction losses, for example) when the switching frequency is high. In this case, even if the converter efficiency remains satisfactory from a system perspective, the local losses in the switching component may prove unacceptable for thermal reasons (component damage).

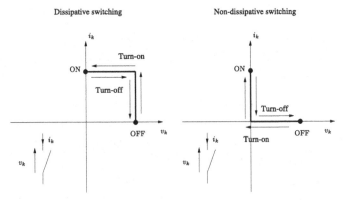

Figure 3.6. *Trajectory of point* (v_k, i_k) *in a switch in dissipative a) and non-dissipative b) switching phases*

In these cases, an auxiliary switching assistance circuit – known as a snubber – may be used. The role of the snubber is not to reduce losses, but to transfer switching losses into a specific resistor. Two distinct types will be discussed in sections 3.2.2 and 3.2.3; these examples are used in the context of a step-down chopper. The first type is used to absorb losses induced by turning on the transistor, while the second deals with losses during the turn-off process. The distinction between dissipative and non-dissipative switching is shown in Figure 3.6, where we see that in the second case, the operating point of the transistor follows the axes of the plane (v_k, i_k) to guarantee an instantaneous power of zero at any moment during the switching process. We therefore need to add elements to produce a transitional voltage drop equal to the power voltage during turn-on phases (see section 3.2.2) and to absorb the load current during turn-off phases (see section 3.2.3). For the purposes of this study, we will consider the step-down chopper diagram presented in Figure 3.7.

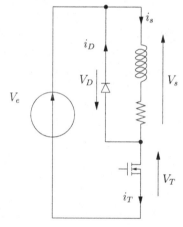

Figure 3.7. *Simple example of a step-down chopper, used in studying transistor switching*

3.2.2. *Reducing turn-on losses*

Turn-on losses in transistors are linked to the fact that the voltage at the terminals remains almost constant while the current increases (and only cancels out when the load current is reached and the free wheel diode turns off). To avoid this problem, a inductor $\mathcal{L}$ may be placed into series with the transistor, which is chosen in such a manner that the voltage at its terminals, of form $\mathcal{L}.di/dt$, is more or less equal to the power voltage V_e of the chopper. Dimensioning in this case is tricky, and detailed observation (initially by simulation and then experimentally) of the evolution of the drain current for a gate control is essential in order to select the appropriate inductor (usually of a low value). For example, the inductance value for a switched current varying between 0 and 10 A in 100 ns, in the case of a power supply V_e of 100 V, is $\mathcal{L} = 1\mu H$.

Once the required inductance value has been calculated, we see that when the inductor is placed into series with the switch, we have a classic problem involving a current source that cannot be turned off. A free wheel diode therefore needs to be associated with this auxiliary inductance, with a resistance R_{da} used to dissipate the stored energy (see Figure 3.8). The resistance is chosen such that the time constant $\tau_{da} = \mathcal{L}/R_{da}$ is negligible in relation to the switching period $T_d = 1/F_d$ and so in terms of power $\mathcal{P}_{da}$. We know that the maximum energy in the inductance is reached when the component is traversed by the load current $i_s = I_0$ (presumed constant):

$$W_{damax} = \frac{1}{2}\mathcal{L}I_0^2 \tag{3.5}$$

The dissipated power $\mathcal{P}_{da}$ is therefore expressed as:

$$\mathcal{P}_{da} = W_{damax}.F_d = \frac{1}{2}\mathcal{L}I_0^2.F_d \tag{3.6}$$

3.2.3. *Reducing turn-off losses*

Turn-off creates losses in a similar way to turn-on, due to the fact that the voltage V_{DS} at the transistor terminals increases to value V_e as soon as the current begins to drop.

These losses may be reduced by the implementation of an auxiliary circuit to reduce the voltage experienced by the transistor during this phase. To do this, we place a capacitor (with capacity C) in parallel to the transistor to limit the increase in voltage at the transistor terminals. In these conditions, the current that would normally circulate in the transistor is drained into the capacitor: the free wheel diode in parallel to the load does not enter into conduction, as the current I_0 continues to circulate in the (transistor + capacitor) assembly – in practice, the capacitor alone, as the current in the transistor cancels out very quickly. The capacitor then charges with a constant current I_0: the voltage at its terminals therefore increases with a mastered slope, I_0/C, but as the transistor is turned off ($i_T = 0$), it does not dissipate switching energy. Finally, when the capacitor is charged at voltage V_e, the parallel free wheel diode enters into conduction.

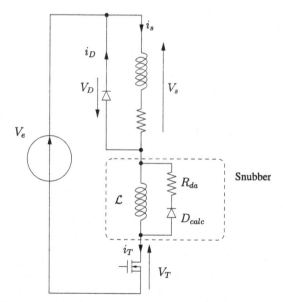

Figure 3.8. *Step-down chopper with turn-on snubber*

A priori the choice of capacitor is not subject to dimensioning constraints; however, it should not be too small, in order to ensure that the transistor will really block before the voltage at the capacitor terminals starts to increase. We may choose a given switching time T_{com} and require $I_0.T_{com}/C$ to remain lower than a few volts, for example (or at least considerably lower than V_e); this value also needs to remain low enough to avoid altering the operation of the chopper in terms of the switching period. The capacitor slows the action of the free wheel diode, and thus leads to an increase in the average voltage supplied to the load (i.e. an increase in the duty cycle). We must therefore ensure that the following inequality is satisfactory:

$$\frac{C.V_e}{I_0} \ll T_d \tag{3.7}$$

For example, we may decide that if $\frac{C.V_e}{I_0}$ is less than 1% of the switching period T_d, the capacitor is correctly dimensioned.

Clearly, in dimensioning a real capacitor, we need to ensure that in addition to the appropriate capacity, the nominal usage voltage V_e of the component must be suitable, and it must be able to withstand a current I_0 during turn-off phases (a pulsed current with a value of I_0 at frequency $F_d = 1/T_d$ and with a narrow width $\beta.T_d$ – e.g. $\beta = 0.01$; in this case, the effective value of the current is $I_0\sqrt{\beta}$).

The capacitor is intended exclusively for use in turn-off, and must therefore be associated with both a diode and a dissipation resistance R_{db} (see Figure 3.9) to allow discharging during conduction phases (with dissipation of stored energy). The dimensioning of the resistance R_{db} involves the same hypotheses used for the turn-on snubber, i.e. a time constant $\tau_{db} = R_{db}C$ that is negligible in relation to the switching period $T_d = 1/F_d$ and a similar power dimensioning $\mathcal{P}_{db}$, which is linked to the energy W_{dbmax} stored in the capacitor:

$$W_{dbmax} = \frac{1}{2}CV_e^2 \qquad \text{[3.8]}$$

and to the switching frequency F_d:

$$\mathcal{P}_{db} = W_{dbmax}.F_d = \frac{1}{2}CV_e^2.F_d \qquad \text{[3.9]}$$

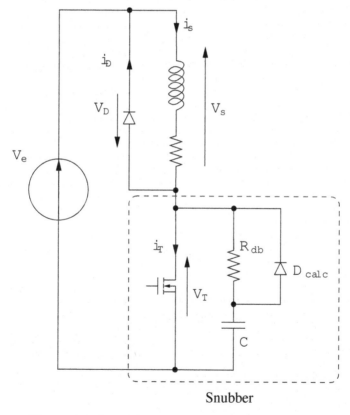

Snubber

Figure 3.9. *Step-down chopper with turn-off snubber*

3.2.4. *Full circuit*

A circuit including elements to reduce losses during both turn-on and turn-off is shown in Figure 3.10.

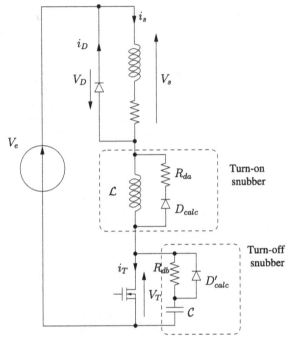

Figure 3.10. *Step-down chopper with full snubber (turn-on + turn-off)*

Note that the snubber circuit only reduces losses in switches, and does not increase the efficiency of the converter. This result comes at a cost, namely an increase in the complexity (and consequently the cost) of the chopper. In applications where a high efficiency is required and with a high switching frequency, justifying the use of a snubber such as that presented above, the current tendency is to use non-dissipative circuits; these belong to a category of converters using "soft" switching, which will be discussed in the following section.

3.3. Firing control of current switches

In this section, we will consider firing control of current switches, i.e.:

– thyristors;

– Triode for Alternating Currents (TRIACs).

Our aim is not to give a full presentation of the different varieties of firing control used in these components, but simply to present the broad operating principles of their trigger circuits and the classic control structures used in both cases. To do this, we look at two widespread components used in this type of equipment:

– pulse transformers;

– Diode for Alternating Currents (DIACs).

This is a useful addition to the controls presented above in the case of transistors, because pulse transformers may also be used for these components.

3.3.1. *Thyristor control*

Let us consider a thyristor which we wish to control using the structure presented in Figure 3.11. From this diagram, we see that the thyristor is equivalent to an association of two transistors (PNP and NPN), with a base/emitter junction clearly visible between the trigger and the cathode.

In this case, the switch is insulated from the control via a transformer, the sole purpose of which is to provide galvanic insulation. As the thyristor control is asymmetrical, a problem arises with transformer magnetization. In fact, we see that when transistor T is on, voltage E is applied to the primary winding of the transformer ($v_1 = E$). As the transformer presents a magnetizing inductance (see Chapter 4), the current i_P increases until the transistor is no longer controlled. Diode D_1 enters into conduction in continuity with the current in the magnetizing inductance, and the voltage applied to the primary winding of the transformer reaches $-V_z < 0$ (if we ignore the voltage drop in D_1): the current i_P decreases, and if this phase lasts for a sufficient period of time, we say that demagnetization is complete ($i_P = 0$)[2].

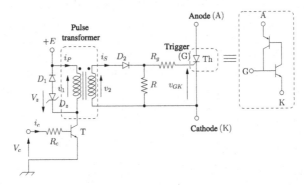

Figure 3.11. *Firing control of a thyristor using a pulse transformer*

The full dimensioning of the assembly depends, first and foremost, on the characteristics of the trigger/cathode junction: in order for the thyristor to fire correctly, the trigger current needs to exceed a certain value, dependent on the temperature. We therefore need to ensure that the component will fire in the worst possible conditions (generally, when cold). For a 2N6397 thyristor (produced by ON semiconductor) with a rating of 400 V/12 A, for example, the maximum trigger current is 30 mA with maximum of 1.5 V. The trigger current must also be maintained for a sufficient period for correct switching to occur: the data are also supplied by the manufacturer, as we see from Figure 3.12, taken from the ON Semiconductor documentation.

---

2 Under normal operating conditions, full demagnetization will occur in this assembly.

From the graph, we see that a pulse of 30 mA with a duration of $T_{\text{pulse}} = 2\mu s$ will lead to turn-on at any operating temperature (including start-up at -40°C). In these conditions, we know that the control will need to drive transistor T for this period, but we still need to dimension the remaining circuit to provide the required current of 30 mA.

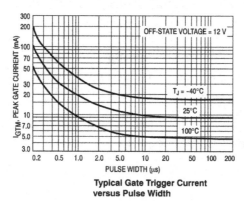

**Typical Gate Trigger Current
versus Pulse Width**

Figure 3.12. *Trigger current as a function of time for three component temperatures (source: ON Semiconductor)*

We will continue to consider the dimensioning of control circuit components using this example, with the assumption that the control signal V_c is of the Transistor-Transistor Logic (TTL) type (0–5 V) and the power voltage E of the circuit is fixed at 15 V. When transistor T is on, the primary voltage v_1 of the transformer may be considered equal to E (leaving aside the voltage $V_{CEsat} \simeq 0.4V$ of T): the magnetizing current (primary) takes the form:

$$i_\mu = \frac{E}{L_\mu}t + I_{\mu 0} \qquad [3.10]$$

If we consider that the circuit operates with full demagnetization, $I_{\mu 0} = 0$. We must therefore ensure that current i_μ does not exceed a certain threshold at the end of a time period T_{pulse}. In practice, manufacture documentation for pulse transformers includes a parameter, denoted as $V.T$

(in volts.seconds – V.s) corresponding to $L_\mu . I_{\mu\,\mathrm{max}} = E.T_{\mathrm{pulse}}$. In our case, given that $E = 15$ V and $T_{\mathrm{pulse}} = 2\mu s$, we must use a transformer with parameter $V.T$ equal to (or greater than) 30 V.μs.

The second parameter used in a transformer is the transformation ratio. This choice is partially arbitrary, and must be adapted based on the values available in manufacturer catalogs (e.g. Schaffner). In this case, we will consider an IT 258 transformer (Schaffner) with a transformation ratio of 1:1 (which easily withstands a current of 30 mA, and gives a $V.T$ product of 250 V.μs).

In these conditions, during power-up of T, the secondary voltage is approximately equal to E, and thus:

$$E = V_{D1} + R_g . I_g + V_{GK} \qquad [3.11]$$

where $V_{D1} = 0.7$ V and $V_{GK} = 1.5$ V max.

From this, we deduce that:

$$R_g = \frac{E - V_{D1} - V_{GK}}{I_g} = 427\Omega. \qquad [3.12]$$

The resistance R has no direct impact on thyristor control[3], except in cases where the transistor is not driven. It contributes to the desensitization of the thyristor to rapid variations in voltage v_{AK}. One drawback of thyristors is the risk of uncontrolled turn-on linked to this brusque variation in the voltage between the anode and cathode (high: $\frac{dv_{AK}}{dt}$). Desensitization is carried out during the manufacturing project, degrading the trigger/cathode junction; this junction no longer behaves as a simple PN junction, and instead behaves as if a low-value parallel resistor was present [LEF 02]. In this case, a resistance of $R = R_g$ would not be

---

3 This value is not always shown on thyristor control diagrams.

problematic, as the secondary winding of the pulse transformer can easily withstand a current that is double of I_g.

NOTE 3.1.– To protect the thyristor from dv_{AK}/dt, we also insert a circuit RC (series) in parallel to the component (between the anode and the cathode). This type of auxiliary circuit, as seen previously for transistors in a chopper, is known as a snubber.

During the turn-off phase of transistor T, conduction through the diode D_1 leads us to apply the voltage $-V_z$ generated by D_z to the primary of the transformer ($v_P = -V_z$). This ensures rapid demagnetization of the transformer (with a duration of T_{dem}). Demagnetization is complete when the surface $V_P.T_{\text{dem}}$ is equal (more precisely, opposite) to $E.T_{\text{pulse}}$:

$$V_P.T_{\text{dem}} = E.T_{\text{pulse}} = V_z.T_{\text{dem}} \qquad [3.13]$$

Consequently, the choice of V_z allows us to fix a demagnetization time that will be the minimum time between two command pulses. With the $V.T$ transformer and V_z parameters of the Zener diode, we can define a maximum duty cycle α_{max} for the control:

$$\alpha_{\text{max}} = \frac{T_{\text{pulse}}}{T_{\text{pulse}} + T_{\text{dem}}} = \frac{V.T/E}{V.T/E + V.T/V_z} = \frac{V_z}{E + V_z} \qquad [3.14]$$

This constitutes a major constraint for the use of a pulse transformer: the control pulses must be short to avoid saturation of the magnetic circuit, and the duty cycle must also be limited. In the case of a thyristor control, this does not pose a real problem, but when controlling transistors over a long period of time, the constraint presents significant problems. These problems are overcome using wave train control. In this case, we may select a voltage V_z equal to E, i.e. 15 V.

NOTE 3.2.– The structure of the control circuit is very similar to that of an insulated switch-mode power supply, as we will see in Volume 3. This is known as a forward power supply due to the fact that energy is transferred directly from the primary to secondary winding (we do not aim to store energy in the transformer, although, in practice, this always occurs). In the case of a classic switch-mode power supply, however, priority is given to the efficiency; magnetization energy is thus retrieved (in a non-dissipative manner) using a third winding in the transformer. In the case in question here, energy performances are not important, and simplicity is key; this energy is therefore dissipated in the Zener diode.

Zener diode dissipates a certain power P_z, and this needs to be taken into account in dimensioning (in addition to V_z). The technical documentation of the transformer gives the primary (i.e. magnetizing) inductance $L_p = L_\mu$, allowing us to calculate the stored energy E_{mag}:

$$E_{\mathrm{mag}} = \frac{1}{2}L_\mu.I_{\mu\,\max}^2 = \frac{1}{2}L_\mu.\left(\frac{E.T_{\mathrm{pulse}}}{L_\mu}\right)^2 = \frac{E^2.T_{\mathrm{pulse}}^2}{2L_\mu} \quad [3.15]$$

where $L_p = L_\mu = 2.5\mathrm{mH}$ for the IT 258 transformer.

Noting the control period $T_d = 1/F_d$ (frequency F_d), we can deduce the power to dissipate in the Zener diode:

$$P_z = \frac{E^2.T_{\mathrm{pulse}}^2}{2L_\mu.T_d} = \frac{E^2.T_{\mathrm{pulse}}^2.F_d}{2L_\mu} \quad [3.16]$$

In the case of a controlled rectifier connected to a 50 Hz network, the thyristor will receive a command once per period, i.e. every 20 ms. We thus have a power of $9\mu\mathrm{W}$ (a very low value that does not require the use of a dissipater, and may be chosen with Surface Mount Device (SMD) package, as in the case of the Rohm EDZTE6115B diode – 15 V/0.15 W).

The rest of the structure dimensioning process consists of selecting a transistor T able to withstand a maximum voltage of $E + V_z = 30V$ and a maximum current $\frac{E.T_{pulse}}{L_\mu} + I_g = 412$ mA. A BD 135 type transistor is perfect in this case, with a rating of 45 V/1.5 A. For diodes D_1 and D_2, components able to withstand maximum currents of 12 and 400 mA, respectively, and voltages of 15 V in both cases are easy to find ("signal" diodes of type 1N4148 would be sufficient).

3.3.2. TRIAC control

For the purposes of TRIAC control, we will consider the simplest structure using an auxiliary component known as a DIAC. The associated symbol and idealized characteristic are shown in Figure 3.13. A DIAC is a dipole with high-impedance behavior for voltages below a certain threshold (symmetrical for positive and negative voltages) that allows current to pass (i.e. presents a low impedance) above this threshold (again, for both positive and negative voltages).

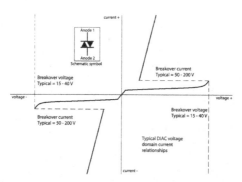

Figure 3.13. *DIAC – symbol and idealized characteristic*

With a component of this type, we may install a rudimentary trigger circuit in order to control a TRIAC in synchronization with the power network of a dimmer: to do this, we simply require a phase shifting circuit to delay the control voltage in relation to the network in order to check the turn-on angle ψ. A proposed control diagram, based on this principle, is shown in Figure 3.14 in the context of a power supply to a lamp.

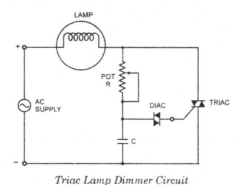

Triac Lamp Dimmer Circuit

Figure 3.14. *Diagram of a single-phase TRIAC dimmer controlled by a DIAC*

4

Passive Components – Technologies and Dimensioning

4.1. Capacitors

4.1.1. *Physical composition and electrostatics*

A capacitor is a passive linear dipole, generally composed of two metal plates connected to the solder points of the casing and separated by a thin insulating layer (dielectric with permittivity ε). For this reason, when a potential difference v_c between the two plates is applied, an electric field appears, and opposing electrical charges ($+q$ and $-q$) are established in the two plates. Mathematically, the simplest capacitor layout to analyze is that of the infinite plane capacitor (Figure 4.1), as it makes easy application of the Gauss theorem to calculate the electric field possible. We will, therefore, consider a set of Cartesian coordinates (O, x, y, z) and a capacitor made up of two flat infinite planes, parallel to the plane (O, y, z) and separated by a gap e (the planes are located on the abscissa $x = -e/2$ and $x = +e/2$). We will consider that these two planes are polarized with a potential difference U (with the highest potential in the plate located at $x = +e/2$).

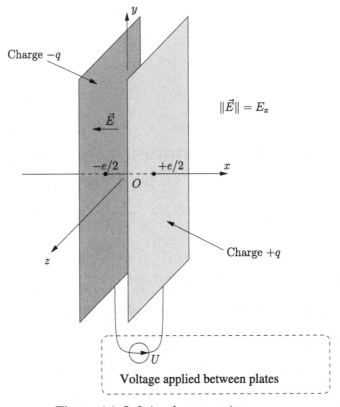

Figure 4.1. *Infinite plane capacitor*

In this situation, we know that an electric field will be generated between the two plates; the geometry of the system (infinite, following axes y and z) means that the electric field $\vec{E}$ (oriented from right to left) will have a single non-null component following the x-axis (of amplitude E_x). Furthermore, we know that the connection between the voltage and the field is an integral, in this case reduced to:

$$U = V(+e/2) - V(-e/2) = \int_{-e/2}^{+e/2} E_x(x).dx. \qquad [4.1]$$

We do not know, *a priori*, whether the electric field between the plates is uniform; so we must bear in mind that $E_x(x)$ may not be constant. However, we may consider that the dielectric layer placed between the plates contains no electrical charge, and so, by applying the Gauss theorem to a straight prism (volume Ω and edge $\partial\Omega$) totally contained in the space between the two plates:

$$\oiint_{\partial\Omega} \overrightarrow{E}.d\overrightarrow{S} = [E_x(x_1) - E_x(x_2)].S$$

$$= \frac{1}{\varepsilon} \cdot \iiint_{\Omega} \rho(x,y,z).dx.dy.dz = 0 \qquad [4.2]$$

Consequently, the values of E_x at positions x_1 and x_2 are identical; therefore, the field between the plates is uniform. Equation [4.1] thus becomes:

$$U = E_x.e \qquad [4.3]$$

By applying the Gauss theorem, we can easily show that the electric field (which must be null at infinity following the x axis) must be null for $|x| > e/2$. This implies the presence of charges in the plates, and we can thus introduce the notion of a surface charge σ such that, by applying the Gauss theorem to a straight prism with axis x, encompassing a section S of a single plate, we obtain the following relationship between σ and E_x:

$$E_x.S = \frac{1}{\varepsilon} \cdot \sigma.S \implies E_x = \frac{\sigma}{\varepsilon} \qquad [4.4]$$

In practice, it is impossible to create an infinite plane capacitor, but a real capacitor may come close to this theoretical case if the section S_a of the plates is large in relation to the distance between them, e. In this case, we can calculate the total charge q in a plate:

$$q = \sigma.S_a = \varepsilon.E_x.S_a \qquad [4.5]$$

We then replace the field E_x by its expression as a function of U to obtain the following well-known relationship:

$$q = \frac{\varepsilon.S_a}{e} \cdot U \qquad [4.6]$$

where we may observe the capacitance $C = \frac{\varepsilon.S_a}{e}$ of the capacitor (in Farads $- F$).

The capacitance of a capacitor is, therefore, affected by three parameters:

– the permittivity of the chosen dielectric material;

– the surface of the plates;

– the thickness of dielectric material between the plates.

Note that the layout of real capacitors is almost never that of a plane capacitor (particularly in power electronics), but the results established above remain valid for initial analysis in all cases.

It is important to remember that the capacitance is not the only important aspect of a capacitor. The choice of a capacitor involves a number of key parameters in addition to the capacitance (again, particularly in power electronics):

– the nature of the voltage at the terminals (alternating or direct);

– the rated voltage;

– the effective value of the current circulating in the capacitor;

– the operating frequency;

– the operating temperature (linked to the environment, as well as to losses in the capacitor itself).

With regard to the last point, it is important to note that a capacitor does not behave like an ideal capacitance, for which we would have a current i_c/voltage v_c relationship of the form:

$$i_c = C \cdot \frac{dv_c}{dt}, \qquad [4.7]$$

but rather as a capacitance with parasite elements, primarily a resistance (leading to losses) known as the equivalent series resistance (ESR). The physical phenomena responsible for this resistance include the ohmic nature of the electrical connections and plates, but also – especially – the loss phenomena in the dielectric layer, which are often modeled using a complex permittivity of the form:

$$\bar{\varepsilon} = \varepsilon_x - j.\varepsilon_y \qquad [4.8]$$

This gives a capacitor impedance $\overline{Z}_C$ of the form:

$$\overline{Z}_C = \frac{1}{jC\omega} = \frac{e}{j.\bar{\varepsilon}.S_a.\omega} = \frac{e}{j.\left(\varepsilon_x - j.\varepsilon_y\right).S_a.\omega}$$

$$= \frac{e}{j.\varepsilon_x.S_a.\omega + \varepsilon_y.S_a.\omega} = \frac{1}{j.C.\omega + \frac{\varepsilon_y.S_a}{e}\omega}, \qquad [4.9]$$

where we clearly see a real component with parallel admittance $\frac{\varepsilon_y.S_a}{e}\omega$. Losses in the dielectric layer can be modeled more naturally by a parallel resistance, corresponding to charge losses in the insulation, but a series model is generally preferred (said, for this reason, to be equivalent, and only given for a certain number of frequencies). Finally, we should also consider the angle of loss δ, a widely used notion in the characterization of dielectrics, which is defined as follows:

$$\tan \delta \triangleq \frac{\varepsilon_y}{\varepsilon_x} \qquad [4.10]$$

4.1.2. *Modeling capacitors*

As we have seen, a capacitor is not solely composed of a capacitance C, and presents a certain number of parasitic components, which may be resumed in an ESR (R_{esr}) and a series inductance L_s. The most widespread model is the one shown in Figure 4.2.

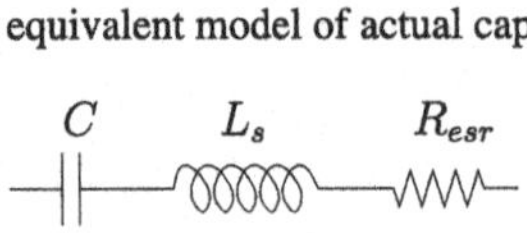

Simple equivalent model of actual capacitors

Figure 4.2. *Equivalent diagram of a real capacitor*

Using a model of this type, we can give an expression of the impedance $\overline{Z}_C$ of the real capacitor:

$$\overline{Z}_C = \frac{1}{jC\omega} + jL_s\omega + R_{esr} = \frac{1 + jR_{esr}C\omega - L_sC\omega^2}{jC\omega} \quad [4.11]$$

and we identify an order 2 numerator for this impedance, which allows us to impose a separate angular frequency ω_0 above which the impedance becomes inductive. This gives a behavior that is completely opposed to the function initially desired. This angular frequency is expressed as:

$$\omega_0 = \frac{1}{\sqrt{L_sC}} \quad [4.12]$$

and thus, for any frequency above $\frac{1}{2\pi\sqrt{L_sC}}$, we need to make sure that no component of the alternating current circulating in the capacitor reaches a very high level. The value of this characteristic angular frequency depends on the value of the capacitance, as well as on the technology used, which has a direct effect on the value of L_s. In practice, generally speaking, electrolytic and large-scale capacitors have lower angular frequencies than those observed in non-polarized or miniature capacitors.

Another important point in capacitor use lies in the associated connections: inductance is essentially due to the internal electrical connections in a component, and we also need to minimize the inductance of external cabling. If we need to place a decoupling capacitor in parallel with a switching cell (e.g. a chopper or an inverter bridge arm), it is important to minimize the length of wire (or traces on a printed circuit) by placing the capacitor as close as possible to the switches.

Finally, to enable capacitive behavior over a wide range of frequencies, a variety of capacitors, often with different technologies, may be used in parallel, in order to combine the advantages (and compensate for the drawbacks) of each individual component. The advantages and disadvantages of different capacitor technologies will be discussed below.

NOTE 4.1.– More elaborate capacitor models can be found in the relevant literature. This lies outside of the scope of this book, but readers may wish to consult [VEN 07] for more information.

4.1.3. *Technologies and uses*

A variety of capacitor technologies exist, differing essentially in the choice of dielectric material. Capacitors and dielectric materials can be divided into two broad categories:

– non-polarized capacitors;

– polarized (or electrolytic) capacitors.

4.1.3.1. *Non-polarized capacitors*

Non-polarized capacitors use plastic or ceramic dielectrics, allowing them to be used with alternating voltages. Some of the most widely used dielectrics include:

– polystyrene;

– polyethylene (or polyester);

– polycarbonate;

– teflon (polytetrafluoroethylene (PTFE));

– ceramics;

– mica;

– polypropylene (PP).

PP capacitors are the most widely used for "pulsed"-type applications, and particularly for power electronics (photograph: Figure 4.3).

Figure 4.3. *A selection of polypropylene capacitors*

These generally have a higher tolerance in terms of capacitance value (i.e. give a less precise value) than polyethylene or polystyrene capacitors, but this is not an issue for this type of application. One significant drawback is linked to the size, as they are bigger than polyethylene capacitors, due to the fact that the material cannot be spread as thinly[1]. The thinnest layer, which is physically achievable with this material, is of the order of a few micrometers $(1 \ \mu m = 10^{-6} m)$.

The main drawback of these components lies in their limited range of operating temperatures. Despite the fact that losses are limited to the electrodes[2], the plastic in question degrades rapidly in the case of overheating (the fusion temperature of PP: 130°C).

Localized destruction of the insulating material leads to self-sealing. When discharge occurs in the dielectric, the metalization at the location disappears [VEN 07], and the surface S_a of the electrode consequently diminishes. This means that the capacitance of the capacitor is also reduced. At the same time, the ESR of the component tends to increase. These two effects combine to degrade the behavior of the capacitor (with an increased impedance for all frequencies).

Ceramic capacitors are another type of non-polarized capacitors used in power electronics. These components use a ceramic dielectric, with different variations according to the target application. Ceramic capacitors are generically known as multilayer ceramic capacitors (MLCCs), and can be divided into three classes:

– Class 1 (based on titanium dioxide, TiO_2) includes components designated by references NP0/CG/C0G which are characterized by low tolerances and excellent stability in

---

1 Capacitance is inversely proportional to the thickness e of the dielectric material. .

2 Unlike the case of electrolytic capacitors, which will be presented in the following section.

terms of temperature. These are used in inverters and in electrical assemblies where precise capacities are required. However, they have a low-volume capacitance and, generally speaking, they are not suitable for decoupling functions in power electronics, particularly in the case of switch-mode power supplies.

– Classes 2 and 3 have high-volume capacities, but the components in these categories are notably less precise than those in class 1 (higher tolerances). They are also extremely unstable in terms of temperature. X7R (class 2) capacitors, for example, present a $\pm 15\%$ variation in capacitance across the range of acceptable operating temperatures (from -55 to +125°C).

Another important point concerning the behavior of MLCC capacitors is their nonlinearity in relation to voltage. The capacitance of an MLCC is highly dependent on the voltage: their nominal value C_0 is measured at very low voltage (less than or equal to 1 V), while the nominal usage voltage may be significantly higher; at this voltage, the effective capacitance may be much lower, as shown in Figure 4.4. As an illustration, an X7R capacitor of 10 μF with a nominal voltage of 400 V presents, in fact, an effective capacitance of around 2 μF if used at 400 V. Care is, therefore, needed in the design of switch-mode power supplies if we wish to use this type of component for filtering. These components are, however, interesting and well suited to use in converters with high switching frequencies (notably interleaved synchronous buck converters used to power microprocessors in personal computer (PC) motherboards). Additionally, this effect is much weaker for these low-voltage applications (e.g. around $V_{core} = 1$ V for an Intel Core i7 processor) than at high voltages. The capacitance drop for a capacitor with a nominal voltage of 25 V does not exceed 20% when this voltage is reached.

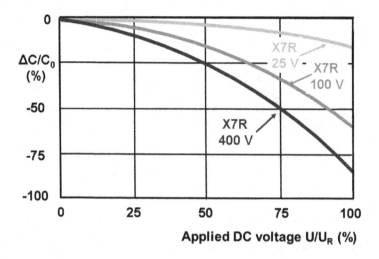

Figure 4.4. *Variation in the capacity of XR7 capacitors as a function of the voltage used for different ratings (25, 100 and 400 V)*

4.1.3.2. *Polarized capacitors*

Polarized capacitors can only be used with a voltage that consistently keeps the same sign. Their potential field of use is, therefore, narrower than that of non-polarized capacitors; however, they present a significantly higher volumetric capacitance. Using this technology, we can, for example, filter the output of a switch-mode power supply in a much smaller space than would be required using a PP component. However, it is important to bear in mind the modeling of real capacitors; the specific frequency of this kind of capacitor is relatively low, and the components themselves are relatively large (electrolytic capacitors are not available in 0805 Surface Mount Device (SMD) casings, for example – see Figure 4.5). Furthermore, the ESR of these capacitors leads to significant damping such that the second-order numerator of $\overline{Z}_C$ (see equation [4.11]) can be factorized into two first-order elements: the capacitor thus behaves in the same way as a resistor across a range of frequencies, of which the natural frequency $f_0 = \frac{\omega_0}{2\pi}$ is the geometric mean. In practice, these capacitors cease to behave in a capacitive manner above a few tens of kilohertz.

Figure 4.5. *SMD polarized capacitors (left) and non-polarized capacitors with 0805 packages (right)*

Two main types of polarized capacitors exist:

– electrolytic capacitors (also known as electrochemical, or simply chemical, capacitors);

– tantalum capacitors.

Electrolytic capacitors consist of two aluminum electrodes in an electrolytic solution. As the anode is oxidized (covered in alumina), continuous current is unable to circulate: this capacitive behavior is reinforced by the electrolytic solution by a "battery" effect (and is not limited to the contribution of alumina as insulation/dielectric). In the case of a polarity reversal, the alumina layer is attacked, and the capacitor finally facilitates the passage of a continuous current leading to a short circuit in the capacitor. Note that in case of burnout (localized alteration of the alumina layer), with the correct polarity, an oxidation reaction will occur naturally and the alumina layer will self-seal (self-repairing capacitor). We may distinguish between liquid electrolyte capacitors and those using solid or gel electrolyte (in the same way as "maintenance free" electrolyte gel batteries) which are particularly suitable for use in power electronics.

Liquid electrolyte capacitors are particularly fragile when used under high stress, as the liquid may evaporate in these cases. Electrolytic capacitors are also subject to faster aging than a large number of other electronic components. They are the main source of failure in electronic components (computer power supplies and domestic equipment such as stereos, televisions, etc.). When a high-capacitance polarized capacitor is required, solid electrolyte technologies are generally preferred for power electronics. These include:

– solid electrolytic aluminum capacitors;

– tantalum capacitors.

Figure 4.6. *Tantalum (left) and aluminum (right) capacitors, equivalent to 22 µF*

Tantalum capacitors offer particularly good performances, but are relatively costly in comparison with aluminum technology. They tend to be used for certain applications due to their higher volumetric capacitance (see Figure 4.6). A summary of the properties of tantalum capacitors is given in Table 4.1 for reference purposes.

However, we should remember that electrolytic aluminum capacitors still account for a large proportion of the polarized capacitors used in power electronics. Returning to the basic capacitance equation $C = \frac{\varepsilon.S_a}{e}$, the main interest of these components does not really lie in the permittivity of the dielectric (alumina has a relative permittivity of around 10), but in the geometric parameters S_a and e. In these cases, the insulating layer is extremely thin compared to plastic film alternatives. This layer is formed by oxidation of the anode during manufacturing, and has a thickness in the order of 1.5 nm/V. Thus, a capacitor with a nominal voltage of 63 V includes a dielectric of a thickness of around 100 nm: consequently, we may hope to obtain a capacitance 10 times higher than that attainable with film for the same surface area. The thinness of the dielectric is not the only advantage of electrolytic aluminum capacitors. The anode is also etched, leading to a considerable increase in the effective surface. In the application guide distributed by Dubillier [COR 13], we see that the surface may be multiplied by a factor of 200 for low-voltage capacitors; this coefficient is lower for higher voltage, but remains significant (up to 60).

Parameter	Result for tantalum capacitors
Normalized service voltage	6.3 to 125 V (reduced above 85°C)
Polarization	Polarized capacitor (reverse voltage resistance limited to a few volts)
Frequency range	Up to around 100 kHz (capacity significantly reduced at higher values)
Current loss	Low: a few μA
Temperature range	-55 to +125°C
Capacity range	From a few μF to a few tens of μF
Tolerance	5% (current value for solid electrolyte capacitors)
Losses	Relatively high (tan $\delta \in [0.05; 0.8]$)
Lifespan/Reliability	Very good for solid electrolyte capacitors

Table 4.1. *Performances of tantalum (solid electrolyte) capacitors*

The drawback of electrolytic capacitors lies in the electrolyte itself. The electrolyte acts as a liquid electrode (cathode), which presents certain advantages; it also has a major drawback. It gives access to the whole surface of the dielectric, including the cavities created by etching, and also regenerates the oxide where necessary (on capacitor assembly – *reformation* – and during the lifetime of the component). However, conduction through an ionic solution is poorer than that obtained in a classic conductor, leading to losses. The ESR of this type of component is, therefore, relatively high (but tends to shrink considerably at higher temperatures); the subsequent heating tends to lead to evaporation of the electrolyte, which escapes from the capacitor through the (imperfect) waterproofing joint created during component assembly.

Electrolyte loss leads to an increase in ESR throughout the component lifetime. This phenomenon is well known, and, in manufacturer data on electrolytic capacitors, the component is considered to have reached the end of its usable life once its ESR has doubled (for a given temperature, generally the nominal usage temperature). However, component aging has little impact on the capacitance, which remains relatively stable throughout the component lifespan (this is not the case for "film" capacitors).

4.2. Inductances

4.2.1. *Physical construction and magnetism*

Fundamentally, an inductance (or self-inductance[3]) is obtained by coiling a conducting wire, potentially around a core made from a magnetic material. As is (too) often the case in electronics, the name "inductance" is used to designate both the physical component (coil) and the corresponding value L (measured in Henrys – H). This value is obtained using the relationship between the electrical values v and i on the one hand, and the magnetic field generated by the coil in the surrounding space on the other hand. We know that the circulation of current in a conducting wire is the source of a magnetic field, based on Ampère's theorem:

$$\oint_{\partial\Sigma} \vec{H}.d\vec{l} = n.I \qquad [4.13]$$

---

3 As opposed to mutual inductance, presented in section 4.3

Moreover, we know that any variation in the magnetic field will induce an f.e.m. e in accordance with *Faraday's law* in the conductor:

$$\frac{dt}{dt}\left(\iint_{\Sigma} \overrightarrow{B}.d\overrightarrow{S}\right) = e \qquad [4.14]$$

Based on these two phenomena, a coil presents a self-inductance behavior, which consists of opposing all variations in the current traveling through it (this behavior is known as *Lenz's law*):

$$v = L\frac{di}{dt} \qquad [4.15]$$

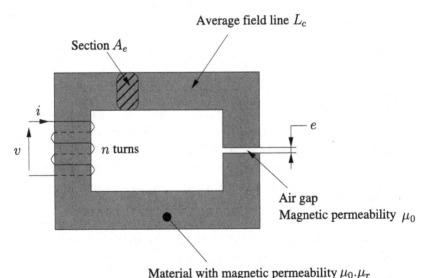

Figure 4.7. *Simplified representation of a coil with magnetic core*

For our applications, it is generally not wise to radiate a magnetic field with a relatively high frequency in the immediate neighborhood of the converters for reasons of electromagnetic compatibility. All of the coils used in electrical engineering and power electronics are, therefore, wound on iron (or ferro-/ferrimagnetic) cores. A simplified representation of a coil of this type is given in Figure 4.7. We note that the magnetic circuit of mean length L_c includes a gap of thickness e, known as the air gap; the role of the air gap will be discussed in the following section.

4.2.2. Modeling a coil with a magnetic core

The iron-core coil shown in Figure 4.7 may be modeled very simply using the following hypotheses:

– linear behavior of the magnetic material with a constant relative permeability $\mu_r \gg 1$;

– a magnetic field that is fully channeled into the circuit (no leaks);

– magnetic flux density B presumed to be uniform in section A_e of the circuit, which can thus be assimilated to a tube with flux $\Psi = B.A_e$ throughout the circuit, with an average length L_c;

– an air gap of length e presumed to be negligible in relation to the dimensions of section A_e.

In these conditions, with a material with high magnetic permeability, we may assume that the lines of the magnetic field traversing the air gap are all parallel to the mean field line, and thus magnetic flux density will be uniform across the air gap and identical to that present in the magnetic material (flux conservation with constant section A_e). Thus, we may apply Ampère's theorem along the mean field line to obtain:

$$H_{fer}.(L_c - e) + H_e.e = n.i \qquad [4.16]$$

where:

– H_{fer} is the magnetic field (or H-field) in the magnetic material (e.g. iron);

– H_e is the magnetic field (or H-field) in the air gap.

The H field and the magnetic flux density B are linked in the following manner:

$$B = \mu.H \qquad [4.17]$$

From this, given that B is constant in the whole of the circuit (iron and air gap), we deduce:

$$\frac{B}{\mu_0.\mu_r}(L_c - e) + \frac{B}{\mu_0}e = n.i \qquad [4.18]$$

Thus, we can write the existing link between B and i, but it is more interesting to calculate the flux Ψ of B through the circuit:

$$\Psi = \iint_{A_e} \overrightarrow{B}.d\overrightarrow{S} = B.A_e = \frac{\mu_0.A_e.n.i}{\frac{1}{\mu_r}(L_c - e) + e} \qquad [4.19]$$

We can also propose an electrical analogy for the magnetic behavior of the circuit, expressing $n.i$ as a function of Ψ: we see that the proportionality coefficient relating Ψ to $n.i$ strongly resembles a resistance R for a conductor of length l, section S and resistivity ρ:

$$R = \rho \cdot \frac{l}{S} \qquad [4.20]$$

This introduces the notion of *reluctance* $\mathcal{R}$. For a magnetic circuit of length l, section S and magnetic permeability μ, this is written

$$\mathcal{R} = \frac{1}{\mu} \cdot \frac{l}{S} \qquad [4.21]$$

Thus, we see that the magnetic permeability is analogous to the inverse of the resistivity (i.e. to the conductivity value) in electrical circuits. This is perfectly coherent with the fact that an iron circuit will channel a magnetic field very efficiently. Furthermore, in our circuit, we see that the flux traverses all of the iron and the air gap, and that the two elements are in a magnetic series. Consequently, the overall reluctance of the circuit is equal to the sum of the reluctances of the two elements:

$$\mathcal{R}_{circ} = \mathcal{R}_{fer} + \mathcal{R}_e = \frac{1}{\mu_0 \cdot \mu_r} \cdot \frac{L_c - e}{A_e} + \frac{1}{\mu_0} \cdot \frac{e}{A_e} \qquad [4.22]$$

This analogy will be useful in understanding the remaining of this chapter. As a final point, note that the magnetic flux may be assimilated to a current, and $n.i$ (the magnetomotive force (mmf) ε) may be assimilated to a voltage (electromotive force (emf)). The reluctance is, therefore, equivalent to a resistance in Ohm's magnetic law:

$$\varepsilon = n.i = \mathcal{R}_{circ}.\Psi \quad . \qquad [4.23]$$

To return to the electrical behavior of the coil, we simply need to apply Faraday's law (for n loops):

$$v = n \cdot \frac{d\Psi}{dt} = \frac{n^2}{\mathcal{R}_{circ}} \cdot \frac{di}{dt} = L \cdot \frac{di}{dt} \qquad [4.24]$$

The inductance of the coil is thus

$$L = \frac{n^2}{\mathcal{R}_{circ}} \qquad [4.25]$$

In our example, the reluctance included two elements (iron and air gap). Noting $\mu_r \gg 1$, we can generally consider that $\mathcal{R}_e \gg R_{fer}$.

To finish, note that the magnetic energy density ω_{mag} is defined by the integral

$$\omega_{mag} = \int H.dB \qquad [4.26]$$

and that, generally, for the whole of the coil, the magnetic energy W_{mag} may be expressed as follows:

$$W_{mag} = \int i.d\Psi = \frac{1}{2}L.i^2 \qquad [4.27]$$

From result $\mathcal{R}_e \gg R_{fer}$, we see that *the purpose of the air gap is to store energy* in magnetic form, while *the iron is not able to do so*: the purpose of the iron is to *channel energy toward the air gap*.

4.2.3. Iron and material losses

The magnetic circuits in coils (or transformers) form the basis of variable magnetic fields, which cause losses known as *iron losses*. These may be split into two distinct groups:

– losses through hysteresis (proportional to the frequency);

– losses through Foucault currents (proportional to the square of the frequency).

These losses essentially depend on the maximum magnetic flux density B_{max} at which the circuit is able to operate, as well as on the waveform of this magnetic flux density for a period (sinusoidal, triangular or other modes). In accordance with Bertotti's model, the expression of losses p_{fer} (in watts) assumes the following form:

$$p_{fer} = k_1.B_{max}^2.f + k_2.B_{max}^2.f^2 \qquad [4.28]$$

Note that this model is fundamentally empirical and is based on a set of parameters k_1 and k_2, which should be adjusted for specific materials and the context in which the magnetic circuit has to be used. We are, therefore, reliant on manufacturer data (or tests, if necessary) to establish a precise model of iron losses.

However, we can still give certain qualitative details concerning the nature of losses and the impact of the chosen materials on these losses. Certain atoms carry a magnetic moment (justified by quantum mechanics), and at macroscopic level, these magnetic moments can induce a magnetic behavior different to that of a vacuum (with magnetic permeability of μ_0, evaluated at $4\pi \times 10^{-7}$ in SI units). Using the notion of relative permeability μ_r, the magnetic permeability μ of a given material is the product $\mu_0.\mu_r$. Based on this consideration, materials may be:

– diamagnetic (with relative permeability μ_r of slightly less than 1, so $\mu < \mu_0$)[4];

– paramagnetic (with $\mu_r = 1$ slightly higher than 1, so $\mu > \mu_0$);

– ferromagnetic (materials with a relative permeability considerably higher than 1). These notably include iron (hence, the category name), nickel and cobalt[5];

– antiferromagnetic (materials containing atoms with potentially high magnetic moments, but with opposition from neighboring atoms, leading to a macroscopic magnetization of zero). The material, therefore, initially appears to present *amagnetic* behavior (like dia- and para-magnetic materials), although these materials behave in a very different manner at the atomic level.

When considering magnetic materials, we need to distinguish between "soft" materials, with high magnetic permeability and zero (or negligible) magnetic flux density with zero H-field, and "hard" materials (*permanent magnets*)

---

4 The most extreme case is that of *supraconductors* that have a magnetic permeability of $\mu = 0$ (for type 1, with H field below the critical field H_c).

5 Other materials of the same type, known as rare earth elements (such as samarium and neodymium), lose their magnetic properties at relatively low temperatures. These materials are notably used in high-performance magnets (SmCo and NdFeBo).

which are saturated, and therefore present a magnetic permeability close to that of a vacuum ($\frac{dB}{dH} \triangleq \mu \simeq \mu_0$) but with a high remanent magnetization (non-null magnetic flux density, which may reach values greater than 1T under zero H-field for certain magnets, such as SmCo and NdFeBo). In this section, we will focus on soft materials, with a specific focus on the ferromagnetic type.

These materials allow us to reach high saturation magnetic flux densities (corresponding to atomic magnetic moments), as their group structure creates magnetic domains separated by "walls" of tens or hundreds of atoms, in which the orientation of magnetization changes. Thus, an iron component is made up of multiple magnetized domains, which produce a macroscopic magnetic flux density of zero (corresponding to the behavior of a soft, i.e. non-magnetized, material). The main problem with ferromagnetic materials is that their electrical conductivity makes them subject to Foucault current in cases with a variable magnetic field (variable magnetic field $\implies$ emf induced in accordance with Faraday's law $\implies$ currents induced by the emf in accordance with Ohm's law $\implies$ joule losses in the material). The first two steps in this sequence of physical phenomena are inevitable, as they do not depend on the material, but on the way in which the magnetic circuit is used. However, the induced currents depend on the conductivity of the material, and the penetration into the material is dependent on the H-field. It is, therefore, important for the material to provide the lowest possible level of conduction, and we may wish to separate the material into insulated elements to produce active sections with low thickness in relation to the thickness of the skin δ_f:

$$\delta_f = \sqrt{\frac{2}{\omega \mu \sigma}} \qquad [4.29]$$

where ω is the angular frequency of the field, μ is the magnetic permeability of the material and σ is its electric conductivity.

Ferromagnetic materials can be used in rolled sheets, offering isotropic magnetic behavior in the sheet plane (sheets with non-oriented grain, for the production of transformers and inductances). These are generally alloyed to produce a good compromise between the following two elements:

– magnetic flux density at high saturation;

– minimized electrical conductivity.

For reference, Table 4.2 shows the relative permeability, maximum operating magnetic flux density[6] and the electrical conductivity of different materials (pure metals and alloys).

Material	Relative permeability μ_r	Magnetic flux density B_{max}	Resistivity $\rho = 1/\sigma$
Fe	1,500 to 10,000	1.2 to 1.6T	$10^{-7}\,\Omega.m$
Ni	600	0.35 to 0.4T	$7 \times 10^{-7}\,\Omega.m$
Co	250	0.4 to 0.5T	$5.8 \times 10^{-7}\,\Omega.m$
Fe-Si (GO)	65,000	2T	$4.8 \times 10^{-7}\,\Omega.m$
Fe-Co	5,000 to 12,000	0.6 to 1.2T	3.5×10^{-7} to $4 \times 10^{-7}\,\Omega.m$
Fe-Ni	6,000 to 220,000	0.8 to 1.6T	3.5×10^{-7} to $6 \times 10^{-7}\,\Omega.m$

Table 4.2. *Magnetic and electrical properties of different base materials (pure metals and alloys)*

An iron–silicon alloy offers a good compromise between magnetic flux density when saturated and electrical conductivity, making it particularly interesting for low-frequency magnetic circuits, especially where cost is an issue: iron and silicon are considerably cheaper than cobalt, used for high-performance magnetic circuits, particularly in the field of aeronautics.

---

6 The limit of linearity which should not be exceeded when producing an inductance. For transformers, it is possible to go slightly above the saturation knee point.

The use of an alloy containing iron and another element (whether magnetically active, such as cobalt or nickel, or inactive, such as silicon) increases resistivity (and thus reduces losses due to Foucault currents). However, the gain is only of factor 5, making the use of these materials difficult when working at frequencies of above or approximately 10 kHz, such as those generally encountered when using switch-mode converters (choppers, resonance inverters, switch-mode power supplies, etc.). In these cases, structured materials are more effective in reducing losses:

– ferrites;

– iron powder.

These materials facilitate a significant increase in electrical resistivity, but this results in a notable reduction in magnetic performance. In the case of ferrites, two magnetic materials are combined, producing imperfect antiferromagnetism (*ferrimagnetism*) as the magnetic moments of the two elements are different. This gives us a non-null (but significantly lower) magnetic moment; these materials are obtained by sintering metal oxide powders, producing apparently homogeneous pieces (very hard, but brittle, and so sensitive to shocks) with very high resistivity (from 1 to $10^8 \Omega$.m). In the case of iron powder, the powder is sintered to produce an agglomerate, with magnetic circuits that are geometrically close to those obtained with ferrites. We obtain a material which is homogeneous in appearance, with a distributed air gap and a reduced magnetic permeability, as the iron grains are insulated from each other, leading to a reduction in losses through Foucault currents. These currents are only able to develop in grains, which are very small (approximately 10 μm, i.e. smaller than the skin thickness at the switching frequencies in usual power electronics converters).

The final group of materials used in switch-mode power supplies is that of amorphous materials (also known as metallic glasses) and nanocrystalline materials. Amorphous materials are metals with an atomic structure equivalent to that found in liquids, i.e. disordered, unlike classic metals, which have a crystalline structure. By definition, a solid with the atomic organization of a liquid is referred to as a glass. The orientation of magnetic domains in these materials, which are initially weakly anisotropic, can be modified after fabrication using magnetic, thermal or mechanical techniques to obtain specific magnetic properties:

– high relative permeability[7] (to obtain a transformer circuit or a magnetic analog saturable inductance for an electric switch);

– low permeability[8] enabling energy to be stored in spite of a total absence of air gaps.

Material	Relative permeability μ_r	Magnetic flux density B_{max}	Resistivity $\rho = 1/\sigma$
Ferrites	7,000	0.4T	$1 \text{to} 10^8\,\Omega.\text{m}$
Iron powder	1,000	1T	insulation

Table 4.3. *Magnetic and electrical properties of ferrites and iron powders*

Geometrically speaking, these materials are relatively limiting, as they assume the form of metallic ribbons. For this reason, they are generally used directly to produce cores (toroids) for coil winding.

From a manufacturing perspective, these materials are obtained by extremely rapid cooling (of the order of 10^3 to 10^6 K/s), achieved by projecting a metal (alloy) in fusion onto a cold wheel, turning at speed, to accelerate cooling. This process is the reason these materials are generally encountered in the form of a ribbon (see Figure 4.8).

---

7 By moving magnetic boundaries.
8 Using a magnetization rotation mechanism.

Considerable progress has been made since the first experiments, in 1960, to obtain metallic glasses that are more stable over time and easier to produce (with slower cooling speeds): these alloys use vitrifying or anticrystallization agents, allowing the metal to be fixed in a solid state with an atomic configuration similar to that of a liquid.

Figure 4.8. *Amorphous ribbon*

Amorphous and nanocrystalline materials have a high magnetic permeability and are extremely thin (meaning they are not particularly sensitive to Foucault currents); these properties make them particularly interesting for applications in power electronics when compared with classic Fe-Si sheets. These materials are notably used for switch-mode power supplies, and particularly in Electro Magnetic Compatibility (EMC) filters.

4.2.4. *Core and winding technologies*

Iron cores are generally made up of FeSi (iron–silicon alloy) grain-oriented electrical steel (GOES) covered with an insulating varnish. They are often produced following E or I layouts, as shown in Figure 4.9.

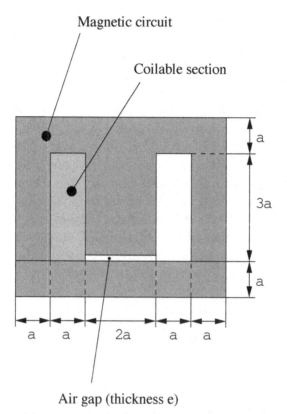

Figure 4.9. *Coil with iron-core in a magnetic circuit using E and I layouts*

Iron–silicon alloys are essentially used to minimize losses due to Foucault currents in iron. Iron is subject to electrical currents induced by variations in the magnetic field, due to the fact that emfs occur (as in conductors) in accordance with Faraday's law; as iron is a conductor, these emfs lead to the emergence of currents. An increase in the resistivity of the material might be thought to be damaging, but we should note that the source is of the "voltage" type, and the losses, therefore, take the form V^2/R.

An increase in the resistivity of the material, therefore, leads to a reduction in losses. This increase in resistivity is obtained by adding a small quantity of silicon to the iron. This does not improve the magnetic behavior of the material, but increases the resistivity. In practice, the quantity of silicon does not exceed 3 to 3.5%, due to the difficulty of working with the alloy, which becomes too brittle above this threshold. Alloys with a higher silicon content (6.5%) can, however, be obtained using very specific production methods (by depositing silicon in gaseous phase). This structure allows us to produce magnetic circuits suitable for the installation of a coil or multiple coils on the central leg of the E, and also leads to optimization of use of a metal sheet with a given surface area, as two couples (E and I) may be taken from the initial rectangle, as shown in Figure 4.10.

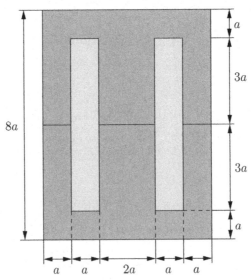

Figure 4.10. *Cross-section of E and I pieces in a magnetic circuit in a rectangular plate*

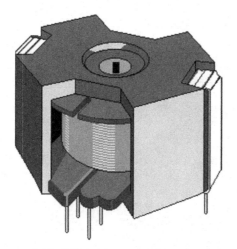

Figure 4.11. *RM ferrite pot with integrated coil, wound on a nylon base*

Ferrite cores may assume one of two forms:

– pots, for wire coils (see Figure 4.11) or copper strap;

– *planar* type pots, for winding on a printed circuit (see Figure 4.12).

4.2.5. *Designing an iron-core filter inductor*

Iron-core coils, such as the one shown in Figure 4.9, are dimensioned using a specification, which must define:

– the required inductance L (in Henrys);

– the maximum operating current I_{max} and its effective value I_{eff} (dependent on the temporal form of the current);

– the wave frequency f of the current;

– where necessary, the intended cost (which affects the choice of sheet quality[9]).

_____

9 Other materials may be used instead of iron–silicon alloy, for example iron–cobalt alloy, which is more costly but offers better performances.

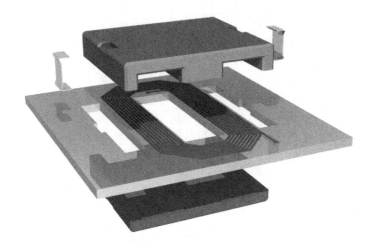

Figure 4.12. *Expanded view of a coil on a printed circuit, inserted into a planar ferrite pot (source: wikipedia)*

A maximum magnetic flux density B_{max} in the material must be fixed at the start of the coil dimensioning process in order to guarantee acceptable (linear) operation. We must also define a maximum current density J_{max}. It is generally best to select a material at the beginning of the process (as this has an influence on the applicable maximum magnetic flux density) in order to obtain a relative permeability value μ_r.

As a starting point, we need to produce a list of dimensioning parameters (assuming that the material for the metal sheet has already been chosen):

– the number of loops n in the coil;

– the single rib a used to dimension the sheet;

– the thickness H of the stack of sheets;

– the width of the air gap e on the central leg of the magnetic circuit.

These four parameters have a direct impact on the reluctance $\mathcal{R}_{circ}$ of the magnetic circuit, as well as on the windable section S_b as shown in Figure 4.9. The reluctance diagram equivalent to the magnetic is shown in Figure 4.13.

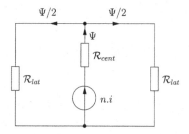

Figure 4.13. *Reluctance diagram of an iron-core coil (E and I)*

The length of the average field line along the central leg is $7a - e$ in the iron section (and e in the air gap); the average field line along each lateral leg is $11a$ in the iron. From this, we are able to deduce the overall reluctance of the circuit:

$$\mathcal{R}_{circ} = \mathcal{R}_{cent} + \frac{\mathcal{R}_{lat}}{2} \qquad [4.30]$$

with

$$\mathcal{R}_{cent} = \frac{1}{\mu_0 \cdot \mu_r} \cdot \frac{7a - e}{2a.H} + \frac{1}{\mu_0} \cdot \frac{e}{2a.H} \qquad [4.31]$$

and

$$\mathcal{R}_{lat} = \frac{1}{\mu_0 \cdot \mu_r} \cdot \frac{11a}{a.H} \qquad [4.32]$$

In practice, we can often use

$$\frac{1}{\mu_0} \cdot \frac{e}{2a.H} \gg \frac{1}{\mu_0 \cdot \mu_r} \cdot \frac{18a - e}{2a.H} \qquad [4.33]$$

and thus write

$$\mathcal{R}_{circ} = \frac{1}{\mu_0} \cdot \frac{e}{2a.H} \qquad [4.34]$$

We see that the 3 geometric parameters of the problem are involved in calculating the reluctance, producing the following expression of the inductance:

$$L = \frac{n^2}{\mathcal{R}_{circ}} = \frac{2\mu_0.a.H.n^2}{e} \qquad [4.35]$$

In this expression, we see the inductance of the iron surface, denoted A_e in section 4.2.2 ($A_e = 2a.H$).

For the "conductor" part of the inductance, we see that the coiled section S_b is equal to $6a^2$ in this structure. We know that n conductors must be contained within this section, and that these conductors must have a cross-section sufficient to make the circulation of an effective current I_{eff} possible, determined so as to remain below a specified current density J_{max}. Moreover, as conductors with a circular cross-section must be placed into a rectangular winding window, a certain amount of space will be lost. This problem must be accounted for using a *winding or filling coefficient K_b* of less than 1 (in practice, between 0.55 and 0.7), reducing the surface that is actually available for conductors:

$$n \cdot \frac{I_{eff}}{J_{max}} \leq 6a^2 K_b \qquad [4.36]$$

We must also consider the second constraint in the dimensioning problem, i.e. the need to respect the maximum magnetic flux density B_{max} defined in the specification. To do this, we use:

$$n.i = \mathcal{R}_{circ}.\Psi = \mathcal{R}_{circ}.B.A_e = \mathcal{R}_{circ}.B.2a.H \qquad [4.37]$$

hence, the constraint:

$$\frac{\mu_0.n.I_{max}}{e} \leq B_{max} \qquad [4.38]$$

In this relationship, we see that the ratio n/e is fixed if we wish to attain the precise specified magnetic flux density:

$$\frac{n}{e} = \frac{B_{max}}{\mu_0.I_{max}} \qquad [4.39]$$

Placing this result back into the inductance expression [4.35], we obtain:

$$L = \frac{2a.H.n.B_{max}}{I_{max}} = \frac{n.B_{max}.A_e}{I_{max}} \qquad [4.40]$$

An effective means of simplifying the dimensioning process (dealing with a number of solutions, with "simple" defined hypotheses and constraints, which may be supposed to be infinite) is to consider, *arbitrarily*, that the thickness H of the stack of sheets must be equal to $2a$ (meaning that the cross-section of the central leg will be square). In this case, we obtain:

$$A_e = 4a^2 \qquad [4.41]$$

and thus:

$$L = \frac{n.B_{max}.4a^2}{I_{max}} \qquad [4.42]$$

If we wish to use the whole of the volume occupied by the coil in an efficient manner ("dense" coil), the coil section needs to tend toward the limit defined in inequality [4.36]:

$$n \cdot \frac{I_{eff}}{J_{max}} = 6a^2 K_b \qquad [4.43]$$

hence,

$$a^2 = \frac{n \cdot I_{eff}}{6.K_b.J_{max}} \qquad [4.44]$$

Substituting this expression into [4.42], we obtain:

$$L = \frac{4n^2 . B_{max} . I_{eff}}{6 . K_b . J_{max} . I_{max}} \tag{4.45}$$

This expression then allows us to select a number of loops n as a function of the desired inductance and the parameters given in the specification. Using equation [4.44], we then calculate the value of a and define the thickness of the metal sheets $H = 2a$.

Once the dimensioning process is complete, we need to select the thickness of individual sheets in accordance with our application. In doing this, we assume that the insulating varnish has a negligible thickness in relation to the thickness of the magnetic material (otherwise, an allowance K_f similar to that of the winding coefficient K_b would have to be used to take account of the reduction in the useful section, A_e); we also need to select a sheet thickness smaller than that of the skin thickness in which Foucault currents may develop, responsible for some of the losses in the magnetic material. In this context, we will assume (without providing a demonstration) that the expression of the skin thickness δ_f as a function of the angular frequency ω of the magnetic field, the magnetic permeability μ and the conductivity σ are as follows:

$$\delta_f = \sqrt{\frac{2}{\omega \mu \sigma}} \tag{4.46}$$

NOTE 4.2.– The skin effect phenomenon should also be taken into account for conductors, in the cases where the current circulating in the inductor involves both large waves and a high frequency. In reality, this is not necessary for iron-core coils operating at frequencies that are generally lower than 1 kHz.

4.3. Coupled transformers and inductors

4.3.1. *Notion of mutual inductance*

Mutual inductance M_{12} is the link that exists between the flux circulating in a coil 1 generated by the circulation of a current i_2 in a coil 2:

$$\Psi_1 = L_1.i_1 + M_{12}.i_2 \qquad [4.47]$$

This is more general than cases of self-inductance, and is essential for understanding the operation of multiple coils on a shared magnetic circuit, such as those used in flyback or forward transformers. One important point when establishing coupled inductance equations is the reciprocal nature of actions between two coils 1 and 2, leading to the equality of mutual inductances (M_{12} and M_{21}, the action of 2 toward 1 and 1 toward 2, respectively):

$$M_{12} = M_{21} \qquad [4.48]$$

In the following section, we will consider the case of two coupled coils in a magnetic circuit with or without an air gap. For this study, as in the case of a single coil, we will use a reluctance diagram model of the magnetic circuit.

4.3.2. *Model of coupled inductors and transformers*

The reluctance diagram in Figure 4.14 gives a fully general model of coupled coils in a magnetic circuit.

Here, we see two elements denoted $\mathcal{R}_{f1}$ and $\mathcal{R}_{f2}$, known as leakage reluctances, associated with coils 1 and 2, respectively. The notion of leakage (which does not really exist for single coils) is extremely important in qualifying magnetic couplings; *a priori*, we wish to maximize the coupling by ensuring that the totality of the flux circulating in coil 1 (e.g. the primary coil of a transformer) also circulates in coil 2 (the secondary coil). In practice, however, part of the flux circulating in coils 1 and 2 is not shared, and constitutes losses due to reluctance. As the circuit involves three loops, three independent fluxes are required to formulate a series of equations; we will consider Ψ_{f1}, Ψ_{f2} and Ψ_{12} (noting that $\Psi_1 = \Psi_{12} + \Psi_{f1}$ and $\Psi_2 = \Psi_{f2} - \Psi_{12}$). Thus:

$$n_1.i_1 = \mathcal{R}_{f1}.\Psi_{f1} + \mathcal{R}_1.(\Psi_{12} + \Psi_{f1}) \qquad [4.49]$$

$$n_2.i_2 = \mathcal{R}_{f2}.\Psi_{f2} + \mathcal{R}_2.(\Psi_{f2} - \Psi_{12}) \qquad [4.50]$$

and

$$n_1.i_1 - n_2.i_2 = \mathcal{R}_1.(\Psi_{12} + \Psi_{f1}) + \mathcal{R}_{12}.\Psi_{12}$$
$$-\mathcal{R}_2.(\Psi_{f2} - \Psi_{12}) \qquad [4.51]$$

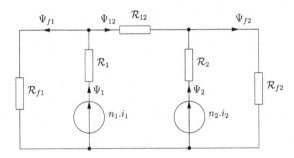

Figure 4.14. *Reluctance diagram of the two-coil circuit*

In matrix form, we obtain:

$$\begin{bmatrix} \Psi_{f1} \\ \Psi_{f2} \\ \Psi_{12} \end{bmatrix} = \underbrace{\begin{bmatrix} \mathcal{R}_{f1} + \mathcal{R}_1 & 0 & \mathcal{R}_1 \\ 0 & \mathcal{R}_{f2} + \mathcal{R}_2 & -\mathcal{R}_2 \\ \mathcal{R}_1 & -\mathcal{R}_2 & \mathcal{R}_1 + \mathcal{R}_{12} + \mathcal{R}_2 \end{bmatrix}^{-1}}_{[\mathfrak{R}]}$$

$$\cdot \begin{bmatrix} n_1.i_1 \\ n_2.i_2 \\ n_1.i_1 - n_2.i_1 \end{bmatrix} \qquad [4.52]$$

The system may easily be reversed; however, the useful values are not the shared flux Ψ_{12} or the leakage fluxes Ψ_{f1} and Ψ_{f2}, but rather the fluxes in the two coils ($\Psi_1 = \Psi_{12} + \Psi_{f1}$ and $\Psi_2 = \Psi_{f2} - \Psi_{12}$). In matrix form, we can write:

$$\begin{bmatrix} \Psi_1 \\ \Psi_2 \end{bmatrix} = \begin{bmatrix} 1 & 0 & 1 \\ 0 & 1 & -1 \end{bmatrix} \cdot \begin{bmatrix} \Psi_{f1} \\ \Psi_{f2} \\ \Psi_{12} \end{bmatrix} \qquad [4.53]$$

We want to obtain a set of equations of form:

$$\begin{bmatrix} n_1.\Psi_1 \\ n_2.\Psi_2 \end{bmatrix} = \begin{bmatrix} L_1 & M_{12} \\ M_{21} & L_2 \end{bmatrix} \begin{bmatrix} i_1 \\ i_2 \end{bmatrix} \qquad [4.54]$$

Based on the previous equations, we may easily deduce that:

$$\begin{bmatrix} L_1 & M_{12} \\ M_{21} & L_2 \end{bmatrix} = \begin{bmatrix} n_1 & 0 \\ 0 & n_2 \end{bmatrix} \cdot \begin{bmatrix} 1 & 0 & 1 \\ 0 & 1 & -1 \end{bmatrix} \cdot [\mathfrak{R}]^{-1} \cdot \begin{bmatrix} n_1 & 0 \\ 0 & n_2 \\ n_1 & -n_2 \end{bmatrix} \qquad [4.55]$$

It is not difficult to verify that the matrix obtained in this way is symmetrical, in accordance with the principle of reciprocal action between the coils: $M_{12} = M_{21}$.

The formulation of a set of equations is greatly simplified by presuming the existence of an ideal coupling (no leakage, with R_{f1} and $\mathcal{R}_{f2}$ tending to infinity) as in this case, the natural (L_1, L_2) and mutual inductances ($M_{12} = M_{21} = M$) can be written as:

$$L_1 = \frac{n_1^2}{\mathcal{R}_1 + \mathcal{R}_{12} + \mathcal{R}_2} \qquad [4.56]$$

$$L_2 = \frac{n_2^2}{\mathcal{R}_1 + \mathcal{R}_{12} + \mathcal{R}_2} \qquad [4.57]$$

$$M = \frac{n_1.n_2}{\mathcal{R}_1 + \mathcal{R}_{12} + \mathcal{R}_2} \qquad [4.58]$$

Transformers are designed based on this hypothesis, but it is important to remember that this objective is never attained in practice. Thus, we will continue the modeling process by considering that the coupling is not ideal and by using any three parameters L_1, L_2 and M. In non-sinusoidal transient mode (leaving aside losses in the conductors), we obtain:

$$v_1 = n_1 \frac{d\Psi_1}{dt} = L_1 \frac{di_1}{dt} + M \frac{di_2}{dt} \qquad [4.59]$$

and

$$v_2 = n_2 \frac{d\Psi_2}{dt} = M \frac{di_1}{dt} + L_2 \frac{di_2}{dt} \qquad [4.60]$$

A modeling approach to show an ideal coupling between the two coils consists of factoring in L_1 in the expression of voltage v_1:

$$v_1 = L_1 \left(\frac{di_1}{dt} + \frac{M}{L_1} \cdot \frac{di_2}{dt} \right) \qquad [4.61]$$

Thus, we introduce the notion of a transformation ratio $m = M/L_1$, noting that, ideally, this ratio should also be equal to the ratio of the number of loops in the two coils (n_2/n_1). We, therefore, note:

$$v_1 = L_1 \frac{d}{dt} \left(i_1 + m.i_2 \right). \qquad [4.62]$$

We can then propose a first diagram equivalent to the primary coil of the transformer, where we see an inductor L_1 traversed by a current which is the sum of the current entering into this first winding and a current proportional to that entering the second winding. This is a coupling phenomenon, which we will attempt to show (although this approach may appear artificial) in equation of the secondary coil. We will begin by writing the expression of $m.v_1$:

$$m.v_1 = \frac{M}{L_1}\left(L_1\frac{di_1}{dt} + M\frac{di_2}{dt}\right) = M\frac{di_1}{dt} + \frac{M^2}{L_1}\cdot\frac{di_2}{dt} \qquad [4.63]$$

This equation includes a term $M\frac{di_1}{dt}$ that is already present in the expression of v_2. Thus, we can rewrite the latter equation as follows:

$$v_2 = m.v_1 - \frac{M^2}{L_1}\cdot\frac{di_2}{dt} + L_2\frac{di_2}{dt}$$

$$= m.v_1 + \left(1 - \frac{M^2}{L_1L_2}\right)L_2\frac{di_2}{dt} \qquad [4.64]$$

In this way, we introduce the notion of the *dispersion coefficient* σ that represents the leaks in the magnetic circuit, defined as follows:

$$\sigma = 1 - \frac{M^2}{L_1L_2} \qquad [4.65]$$

Coefficient σ should be as low as possible; in the ideal case, it is zero, as:

$$\frac{M^2}{L_1L_2} = 1 \qquad [4.66]$$

In practice, however, the inductance $\sigma.L_2$ always exists in equation [4.64]. The model of the transformer can therefore be reduced to three parameters:

– the magnetizing inductance L_1;

– the transformation ratio m;

– the leakage inductance $\sigma.L_2$ (totaled for the second coil, in this case).

The equivalent diagram corresponding to the set of equations is shown in Figure 4.15.

$$\begin{cases} v_1 = L_1 \frac{d}{dt}\left(i_1 + m.i_2\right) \\ v_2 = m.v_1 + \sigma.L_2 \frac{di_2}{dt} \end{cases} \qquad [4.67]$$

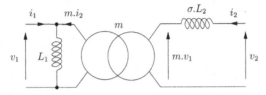

Figure 4.15. *Equivalent electrical diagram of a transformer*

We must also consider one final important aspect of coil coupling in magnetic circuits: coil orientation, which clearly has an effect on the orientation of the fluxes generated as a function of the orientation of currents. To simplify diagrammatic representations of coupled coils, we add a dot (*polarity dot*) or an asterisk to the symbols of each coil, used to orient the e.m.fs generated by mutual inductances: when a current i_k enters a coil k via the equivalent point, it generates an e.m.f. of the form $M_{jk}.\frac{di_k}{dt}$ *oriented toward the polarity dot* of coil j.

In geometric terms, the identification of equivalent points is simplified in cases where the coils share an axis. If, when traveling from one terminal to another, the wire turns in the same direction on both coils, then the equivalent points for the two coils will be the two initial (or two final) terminals. If the winding directions are opposite, then the equivalent points will be the initial terminal on one coil and the final terminal on the other coil. If the two coils are not on the same axis, we need to (virtually) move one of the coils onto the axis of the other in order to apply this reasoning. We will only consider cases with two coils here: in cases with more coils, we must simply select one coil as the reference point and characterize all of the other coils in relation to the first coil. The problem can, therefore, always be reduced to a case of two coils.

5

Designing Printed Power Circuits

5.1. Classic printed circuits

5.1.1. *General and specific points for power electronics*

A variety of technologies may be used for printed circuits, notably in the domain of power electronics. The most traditional approach involves a glass fiber board impregnated with epoxy resin. A circuit may include conducting traces on:

– one side: single-sided printed circuit board;

– both sides: double-sided printed circuit board;

– both sides, plus internal layers: multi-layer printed circuit board.

The most widespread dielectric used in Printed Circuit Board (PCBs) is FR-4 (FR for Flame Retardant). This material is used for cost reasons, as other materials are more expensive and generally used for radio frequency applications (e.g. Rogers RT/duroid, XT/duroid, etc.).

These elements are not specific to power electronics, and the same materials are used for electronic signal or information processing applications. We will discuss materials used specifically for power applications in section 5.2; these essentially promote the evacuation of heat generated by losses in components. Note, however, that the main priorities in designing PCBs for electronic power converters are:

– the ability of the copper traces to withstand current;

– the voltage which the insulation between these traces is able to withstand.

These two current and voltage constraints[1] can (and should) be satisfied independent of the medium used for the PCB.

5.1.2. Trace dimensioning

It is important to be aware of the thickness of copper used in order to dimension traces correctly in accordance with the current used. Charts are available which show the connection between key dimensioning parameters:

– thickness of copper;

– track width;

– intended current;

– heat tolerance of the track.

An example is presented in Figure 5.1 (for an FR-4 type medium). These charts show 3 standard thicknesses of copper (35 μm, 70 μm and 105 μm). The first value (17.5 μm) given is that which is often available at the beginning of the manufacturing process for a multilayer (or at least double sided) PCB, involving plating through holes (PTH). In this case, the holes in the board are coated with copper, and the copper traces on the external layers of the printed circuit are coated at the same time. Thus, a layer of copper with a thickness of 35μm is often obtained by thickening a copper

---

1 One or both of which is generally high in the context of power applications.

layer with an initial thickness of 17 or 18 μm. Note, moreover, that in these conditions, the thickness is simply a nominal value, as we are dealing with chemical, and consequently non-uniform, growth. Manufacturers monitor this process as far as possible in order to obtain a minimum thickness, but this is never constant across the whole surface of a printed circuit. Figure 5.2 shows an image from a manufacturing file control program for PCBs (GERBER). The varying levels of metalization of different zones are shown by shading (the denser the initial coverage of a zone, the less copper will be deposited; inversely, the sparser the initial coverage, the more copper will be deposited).

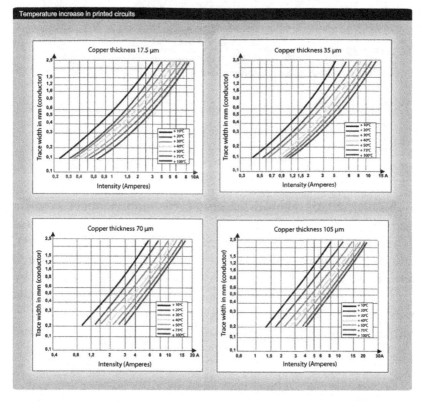

Figure 5.1. *Charts showing the heating of PCB traces For a color version of the figure, see www.iste.co.uk/patin/power1.zip*

Top plating index: 0.32

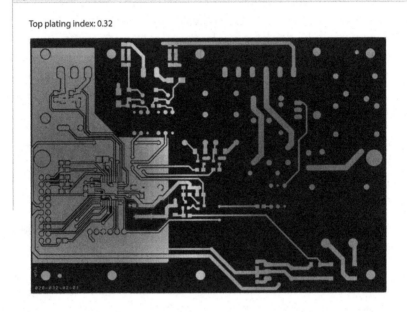

Underplating   Normal plating   Overplating

Figure 5.2. *Evaluation of chemical copper deposits for manufacturing (source: Eurocircuits) For a color version of the figure, see www.iste.co.uk / patin / power1.zip*

Finally, PCBs should be protected against short circuits and overcurrents. Electronic protection mechanisms are generally the fastest (as currents are monitored by sensors), followed by high speed fuses[2], used to protect electronic power converters. To prevent a PCB from being damaged by excessive current, we must ensure that the PCB itself does not act as a fuse. A formula can be used to establish a connection between intensity and time (in the style of a curve

---

2 The main function of these fuses is to avoid fire breaking out, rather than to protect electronic components, which have a relatively low chance of surviving a fault if a close control device does not act sooner (detection of overcurrent or desaturation of a transistor following a short circuit).

in the (I, t) plane of a fuse): this is Onderdonk's formula which, in a simple form applicable to copper traces on a PCB, is written as:

$$I[A] = \frac{0.188A[\text{mils}^2]}{\sqrt{t[s]}} \qquad [5.1]$$

where A is the section of the conductor in mils^2 (i.e. thousandths of inches squared; imperial units are still predominant in electronics). Using metric units, for example with A expressed in mm^2, we obtain:

$$I[A] = \frac{291.4A[\text{mm}^2]}{\sqrt{t[s]}} \qquad [5.2]$$

As an illustration of this point, a conductor with a thickness of 35 μm needing to withstand 20 A for 5 s will need to have a width greater than or equal to 4.4 mm.

NOTE 5.1.– This formula only produces good results for $t \leq 10$ s. We should also note that as time tends toward infinity, the acceptable current tends asymptotically toward 0. The starting hypotheses used in this equation include the fact that heat will remain confined within the conductor, the temperature of which will thus increase indefinitely.

5.1.3. *Insulation between traces*

Electronic power converters need to deal with high voltages, so the design of dense PCBs must not detract from the galvanic insulation between copper traces; a track spacing value must be respected. Creepage may also be used on the printed circuit in order to improve the insulation provided by a given gap.

The first stage consists of determining the creepage resistance index of the material used for the PCB. This parameter is generally specified by manufactures in technical documentation, based on a test protocol imposed by the IEC 60112 standard. In the case of FR-4, this parameter is set at 250 V. Materials may be grouped into 4 classes, as shown in Table 5.1, by Comparative Tracking Index (CTI).

Classes of materials	CTI (in V)
Group I materials	CTI > 600
Group II materials	400 < CTI < 600
Group IIIa materials	175 < CTI < 400
Group IIIb materials	100 < CTI < 175

Table 5.1. *Classification of materials by CTI*

FR-4 falls into Group IIIa in this classification. In cases where the material in question is unknown, it is always assumed to be part of Group IIIb.

Based on this classification, tables may be used to give the spacing needed between two conductors for a given voltage, for different levels of pollution on the surface of the material (from level 1 in the best cases to level 3 for the worst). In practice, for "normal" domestic or industrial applications, we use pollution level 2. Table 5.2 provides full data (concerning materials and pollution levels) for voltages from 10 V to 63 kV.

For voltage values which lie between those specified in the table, linear interpolation can be carried out between the closest neighboring points. The value of the creepage calculated on this basis should then be rounded up to the nearest tenth of a millimeter.

The indicated values guarantee *basic functional insulation*. In cases where *reinforced insulation* is required, we double the obtained length.

Max. effective voltage	Printed circuits		Other materials						
	Degree of pollution 1	Degree of pollution 2	Degree of pollution 1	Degree of pollution 2			Degree of pollution 3		
	I, II, IIIa, IIIb	I, II, IIIa	I, II, IIIa, IIIb	I	II	IIIa, IIIb	I	II	IIIa, IIIb
10 V	0.025	0.04	0.08	0.4			1		
12.5 V	0.025	0.04	0.09	0.42			1.05		
16 V	0.025	0.04	0.1	0.45			1.1		
20 V	0.025	0.04	0.11	0.48			1.2		
25 V	0.025	0.04	0.125	0.5			1.25		
32 V	0.025	0.04	0.14	0.53			1.3		
40 V	0.025	0.04	0.16	0.56	0.8	1.1	1.4	1.6	1.8
50 V	0.025	0.04	0.18	0.6	0.85	1.2	1.5	1.7	1.9
63 V	0.04	0.063	0.2	0.63	0.9	1.25	1.6	1.8	2
80 V	0.063	0.1	0.22	0.67	0.9	1.3	1.7	1.9	2.1
100 V	0.1	0.16	0.25	0.71	1	1.4	1.8	2	2.2
125 V	0.16	0.25	0.28	0.75	1.05	1.5	1.9	2.1	2.4
160 V	0.25	0.4	0.32	0.8	1.1	1.6	2	2.2	2.5
200 V	0.4	0.63	0.42	1	1.4	2	2.5	2.8	3.2
250 V	0.56	1	0.56	1.25	1.8	2.5	3.2	3.6	4
320 V	0.75	1	0.75	1.6	2.2	3.2	4	4.5	5
400 V	1	2	1	2	2.8	4	5	5.6	6.3
500 V	1.3	2.5	1.3	2.5	3.6	5	6.3	7.1	8
630 V	1.8	3.2	1.8	3.2	4.5	6.3	8	9	10
800 V	2.4	4	2.4	4	5.6	8	10	11	12.5
1 kV	3.2	5	3.2	5	7.1	10	12.5	14	16
1.25 kV			4.2	6.3	9	12.5	16	18	20
1.6 kV			5.6	8	11	16	20	22	25
2 kV			7.5	10	14	20	25	28	32
2.5 kV			10	12.5	16	22	32	36	40
3.2 kV			12.5	16	20	28	40	45	50
4 kV			16	20	25	36	50	56	63
5 kV			20	25	32	45	63	71	80
6.3 kV			25	32	40	56	80	90	100
8 kV			32	40	56	80	100	110	125
10 kV			40	50	71	100	125	140	160
12.5 kV			50	63	90	125			
16 kV			63	80	110	160			
20 kV			80	100	140	200			
25 kV			100	125	180	250			
32 kV			125	160	220	320			
40 kV			160	200	280	400			
50 kV			200	250	360	500			
63 kV			250	320	450	600			

Table 5.2. *Minimum creepage distances in mm (in accordance with standard EN 60950-1)*

EXAMPLE 5.1.– If we wish to guarantee reinforced insulation for an effective voltage of 230 V through an FR-4 printed circuit, we should begin by noting that the material falls into Group IIIa. In a domestic or industrial environment, we presume that the degree of pollution is 2. According to the table, the minimum creepage is 0.63 and 1 mm for 200 and 250 V respectively. Linear interpolation between these two values gives us a measurement of 0.852 mm. We will therefore consider a functional distance of 0.9 mm. Reinforced insulation will therefore correspond to 1.8 mm. To obtain this result, we used Column IIIa for PCBs. For another, less suitable material (still within Group IIIa), we would obtain an insulation value of 2.3 mm, with reinforced insulation at 4.6 mm.

These values appear excessive in practice. Generally, to calculate correct spacing, we limit the voltage per unit of distance (value homogeneous to an electric field) to 1.6 kV/mm or 40 V/mil (1 mil = 1/1,000 inch[3]).

The EN 60950-1 safety standard is only applicable to certain types of products. For products not covered by this standard, the IPC (Association Connecting Electronics Industries) has published recommendations in a variety of documents, such as IPC-2221 or the more recent IPC-9592. The latter document specifies a spacing of SPACING (mm) as a function of the "peak" voltage Vpeak between the conductors:

$$\text{SPACING (mm)} = 0.6 + \text{Vpeak} \times 0.005 \qquad [5.3]$$

---

3 1 inch = 25.4 mm (so 1 mm = approx. 39 mil).

The latest updates to this document enable this value to be corrected for voltages below 100 V:

$$\text{SPACING (mm)} = \begin{cases} 0.13 \text{ if Vpeak} < 15V \\ 0.25 \text{ if } 15V \leq \text{Vpeak} < 30V \\ 0.1 + \text{Vpeak} \times 0.001 \text{ if } 30V \\ \leq \text{Vpeak} < 100V \end{cases} \quad [5.4]$$

VERIFICATION 5.1.– We can use this method to calculate the insulation required under the same conditions used in the previous example. For an effective sinusoidal voltage of 230 V, we have a "peak" value of 325 V, giving us an insulation measurement of 2.23 mm. In our previous calculations, we obtained values of 1.8 mm, with 4.6 mm for reinforced insulation. Generally speaking, we should be prudent if insulation is the primary design consideration, but this reinforced insulation will increase the size of the PCB. Moreover, standards linked to target applications are always based on empirical rules. Free programs, such as the Saturn PCB Toolkit (available to download at http://www.saturnpcb.com/ pcbtoolkit.htm[4]) offer trace and spacing dimensioning tools in accordance with the IPC recommendations.

To improve the insulation between traces, we may not only space conductors on the epoxy base, but also create grooves in the insulating material, as air is a better insulator. Figure 5.3 shows a comparison between clearance and creepage; in the latter case, an electrical arc will be forced to go around the groove, rather than traveling across it. This type of insulation is widely used in optocouplers for galvanic insulation of close commands in power transistors at high voltages (IGBT). This solution offers gains in terms of size, and is key to the

---

4 Saturn PCB specializes in the production of printed circuits and component installation, and has released a free PCB design toolkit for use by its clients.

production of miniature integrated circuits. It has been widely used in DIP/DIL type traverse circuits for some time, and is now used systematically for surface mounted components (e.g. SOICs).

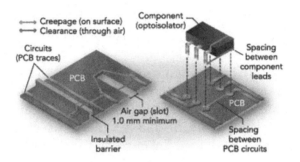

Figure 5.3. *Creepage and clearance in printed circuits for electrical trace insulation*

This data can be contradictory, and does not provide us with a clear and precise idea of rules to follow. A table provided on the Gold Phoenix PCB manufacturing website (http://www.goldphoenixpcb.com/html/SupportResource/prdt/arc125.html) is both clear and precise, and, moreover, perfectly adapted to the requirements of bare board circuit designers. It specifies the insulation required for internal and external layers of a circuit (with or without varnish, and also at altitude[5]). Information is also provided regarding assemblies, i.e. what happens once components are placed onto the PCB. This data is reproduced in Table 5.3.

5.2. Alternatives to the classic PCB

Printed circuits may be produced by stacking layers of conductors and insulating sheets (epoxy resin and/or prepeg

---

5 At high altitudes, where the atmosphere is thinner, the dielectric rigidity of air is reduced, which may seem counterintuitive.

sheets[6]), but other methods may be used, incorporating metallic cores (see Figure 5.4) which enable efficient evacuation of heat.

| | Bare board | | | | | | | | | | | Assembly | | | | |
| | Internal layers | | External conductors (uncoated) | | External conductors (uncoated, altitude >3050m) | | External conductors (coated) | | External conductors with conformal coating | | External component leads (uncoated) | | External component leads with conformal coating | |
Vpk	mm	inch	mm	inch	mm	inch	mm	inch	mm	inch	mm	inch	mm	inch
15 V	0.05	0.002	0.1	0.004	0.1	0.004	0.05	0.002	0.13	0.006	0.13	0.006	0.13	0.006
30 V	0.05	0.002	0.1	0.004	0.1	0.004	0.05	0.002	0.13	0.006	0.25	0.01	0.13	0.006
50 V	0.1	0.004	0.6	0.024	0.6	0.024	0.13	0.006	0.13	0.006	0.4	0.016	0.13	0.006
100 V	0.1	0.004	0.6	0.024	1.5	0.06	0.13	0.006	0.13	0.006	0.5	0.02	0.13	0.006
150 V	0.2	0.008	0.6	0.024	3.2	0.4	0.4	0.016	0.4	0.016	0.8	0.032	0.4	0.016
170 V	0.2	0.008	1.25	0.05	3.2	0.4	0.4	0.016	0.4	0.016	0.8	0.032	0.4	0.016
250 V	0.2	0.008	1.25	0.05	6.4	0.26	0.4	0.016	0.4	0.016	0.8	0.032	0.4	0.016
300 V	0.2	0.008	1.25	0.05	12.5	0.5	0.4	0.016	0.4	0.016	0.8	0.032	0.4	0.016
500 V	0.25	0.01	2.5	0.1	12.5	0.5	0.8	0.032	0.8	0.032	1.5	0.06	0.8	0.032
1 kV	1.5	0.06	5	0.2	25	0.99	2.33	0.092	2.33	0.092	3.03	0.12	2.33	0.092
2 kV	4	0.158	10	0.4	50	1.97	5.38	0.22	5.38	0.22	6.08	0.24	5.38	0.22
3 kV	6.5	0.256	15	0.6	75	2.96	8.43	0.34	8.43	0.34	9.13	0.36	8.43	0.34
4 kV	9	0.355	20	0.79	100	3.94	11.48	0.46	11.48	0.46	12.18	0.48	11.48	0.46
5 kV	11.5	0.453	25	0.99	125	4.93	14.53	0.58	14.53	0.58	15.23	0.6	14.53	0.58

Table 5.3. *Insulation proposals for bare boards and assemblies (source: Gold Phoenix PCB)*

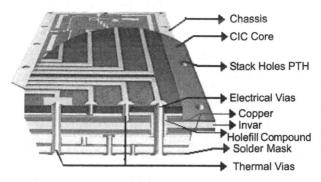

Figure 5.4. *Metal core PCB (source: www.emeraldinsight.com)*

---

6 "Prepegs" are pre-impregnated glass fibers, which are then polymerized in a heated press during the manufacture of multi-layer printed circuits.

A major problem in the design of these stacks lies in the choice of materials. When the system is subjected to raised temperatures, warping can occur due to differing dilation coefficients (note the use of invar in the center of the stack[7]).

Another similar solution consists of using a metallic substrate, onto which an insulating layer is placed, followed by a copper sheet used to produce a classic (single-sided) PCB. The interest of this structure lies in its capacity to evacuate the heat dissipated by components placed onto the printed circuit toward the metallic substrate, through the insulating layer (IMS: Insulated Metal Substrate). The substrate is then generally fixed to a heat sink in order to increase the surface exposed to the air. This circuit offers a solution equivalent to that used in power modules, such as those for IGBT bridge arms or the SKiiP 12AC126V1 module. The main difference lies in the ease of use of this technology, which is suitable for classic SMD type electronic components, whereas power modules use bare chips connected to each other by bonding wires (see Figure 5.5). The associated manufacturing process are therefore quite different, and components may also be integrated.

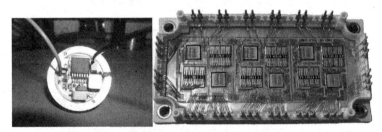

Figure 5.5. *Comparison between insulated metal substrate PCBs (IMS), left, and power modules, right*

---

7 Invar is an iron (64%) – nickel (36%) alloy known for its low dilation coefficient; its name is taken from the word *invariable*.

5.3. PCB assembly

5.3.1. *Connecting components*

Components may be connected to a PCB in different ways, both in electronics as a whole and, more specifically, in power electronics. These methods fall into two broad categories:

– pressure/insertion methods;

– soldering methods.

In the case of pressure connections, spring contacts are pressed using a given force onto tabs on the PCB to create a strong electrical contact. This has already been discussed in Chapter 2, with a module presented in Figure 2.6. The advantages of this connection type include simplicity and robustness. The connections can withstand difficult operating conditions, with high temperatures and/or significant vibrations, with no risk of deterioration. Care is needed to ensure a strong contact is established and that the surface state is satisfactory; the contact point must be free from contamination (dust etc.) and oxidation must be avoided. An Electroless Nickel Immersion Gold (ENIG) finish may be used for high quality PCBs in order to avoid this final issue.

"Pressure" type methods also include nut/bolt type connections. Once again, a good surface state is essential and an ENIG finish is ideal. Note that screw connections for power modules must be carried out with a specified torque in order to guarantee satisfactory electrical contact: this value may be provided in technical documentation, as in the case of screw fixations attaching the base of these components to a heat sink. The problem is the same in both cases, with a main aim of minimizing resistance in the contact point (electrical resistance in the first case, thermal resistance in the second). If performances need to be guaranteed, this type of assembly must be carried out using a correctly-calibrated dynamometric wrench (or screwdriver).

Although these techniques are interesting, they are not widely used in electronics, and soldering remains the norm, including in the specific case of power electronics.

5.3.2. *Soldering*

Component soldering consists of connecting the PCB and the pins of a component using a filler metal, brought to fusion point, with an affinity for the materials used in the two connecting elements. In electronics, the classic filler metal is a tin (Sn – 60%) and lead (Pb – 40%) alloy, which has a relatively low fusion point (between 179 and 183°C). This property is important to avoid placing excessive strain on fragile electronic components during assembly. However, since 2006, the use of lead in electronic equipment is prohibited (RoHS *Reduction of Hazardous Substances* standard) and a new tin (Sn – 96.5%), silver (Ag – 3%) and copper (Cu – 0.5%) alloy is now used. Note, however, that this new alloy poses certain problems:

– higher fusion temperature than lead alloys (around 220°C) leading to higher energy consumption in producing PCBs, and to greater stress on components;

– the creation of whiskers[8] between component pins during use, leading to short circuits.

Note that the difference between lead-based and lead-free soldering is relatively easy to see:

– lead alloys have a shiny appearance after soldering, whereas lead-free alloys are matte;

---

8 Tin filaments, formed when a voltage is applied between two neighboring pins of a component. When filaments touch, a short circuit is established. This phenomenon is particularly critical in modern SMD components due to the low pitch between pins.

– test kits (see Figure 5.6) may also be used to remove any doubts regarding the nature of an alloy (using a reagent which turns pink when exposed to lead).

Figure 5.6. *RoHS conformity test (Techspray Instant Lead Testing Swabs 2515)*

Two types of industrial manufacturing process are used:

– wave soldering, which is particularly useful for through-hole assembly, but may also be used for SMD components;

– reflow soldering (in an oven), which is best suited to SMD components but may also be used for through-hole components.

Figure 5.7. *Photograph of a wave (source: http://www.orion-industry.com/questions.html)*

5.3.2.1. *Wave soldering*

Wave soldering (Figure 5.7) consists of using a bath of liquid solder (tin alloy) in which a wave is created, establishing contact with the printed circuit, carrying the components for connection (through hole and/or SMD). This technique is particularly fast, and is therefore suited to use with large series (see Figure 5.8). However, the use of this technique must be planned from the beginning of the PCB design phase, as component placing is important: the wave deposes the alloy in a non-symmetrical manner according to the placement of contact points ("upstream" or "downstream" from a component in relation to the wave). In these conditions, there is always a risk of solder accumulation, and despite the use of solder mask[9] (generally green on classic PCBs), short circuits can occur between neighboring components, particularly in cases with a high density of miniature components.

Figure 5.8. *Wave soldering machine (ERSA ETS-330)*

5.3.2.2. *Reflow soldering*

The second technique used in industry for soldering electronic components involves the use of a reflow oven. This process involves several steps:

– a solder paste (a mixture of powdered solder and tinning flux[10]) is applied to the relevant contact pads using a stencil;

---

9 The role of solder mask is to prevent the tin alloy from adhering anywhere but the contact points.

10 The role of the flux is to deoxidize the pieces being soldered, prevent oxidation during soldering (which occurs more rapidly the higher the temperature), to facilitate wettability of the elements for connection and to improve the adhesion of the solder.

– the component is placed on the PCB;

– the PCB and components are placed into a reflow oven.

The machines used in this soldering technique are shown in Figure 5.9.

Figure 5.9. *Paste application, component placing and reflow (Manncorp AP3000, MC-388, CR-4000C)*

Note that these steps can all be carried out in a laboratory context for prototyping purposes with less sophisticated equipment (see Figure 5.10).

Figure 5.10. *Paste stenciling (Eurocircuit eC-Stencil-Mate), component placing (LPKF Protoplace S) and reflow (LPKF Protoflow S)*

5.3.3. *High temperature brazing materials*

As we saw in Chapter 2, modern components for power electronics can be used at increasingly high temperatures. At 200°C, classic lead soldering cannot be used, and the use of the standard alloy (Sn-Ag-Cu) is also questionable, due to its relatively low fusion temperature. Conformity to the RoHS standard is problematic; whilst 30% lead solder alloy has a low fusion temperature, classic High Melting Point (HMP) alloys also contain lead, and in greater quantities (Pb 93.5%, Sn 5%, Ag 1.5%). The fusion point of these alloys is around 300°C. Substitutes exist, but significant work is still underway in the domain, as the long-term behavior of these alloys is not well-understood (aging). Moreover, certain alloys are particularly expensive: one notable example is the 80Au/20Sn alloy, which, as its name indicates, contains 80% gold, and gives a fusion temperature of 280°C. Bismuth (Bi) seems to offer promising prospects for the creation of ecologically-sound, economical and reliable solutions. This is an essential aspect of high temperature electronics, particularly in allowing us to exploit the full potential of SiC components.

5.4. Additional information

Further details on PCB design may be found in Volume 4, concerning electromagnetic compatibility, and more specifically in Chapter 2. In this chapter, special consideration will be given to impedance coupling problems, with the notion of mass planes and stitching vias, and the diaphony involved in capacitive and inductive couplings between traces. Generally speaking, the process of designing an electric chip can be complex, as the designer must take account of a number of multi-physical constraints:

– electrical constraints, evidently, in order to ensure that the electrical connections between components will withstand the relevant voltages and currents;

– electromagnetic constraints, to avoid interference between the chip and its immediate environment, and between the subsystems contained in the chip itself;

– mechanical constraints, including size and the mechanical connections between components, the chip and the medium (casings, racks, etc.);

– thermal constraints (taking account of the fragilities of different components and the ability to evacuate heat, notably using air flows).

Figure 5.11. *3D view of an inverter bridge arm integrated into a rackable 3U casing (produced by UTC using Altium Designer)*

Modern Electronic Design Automation (EDA) software offers increasing levels of assistance to engineers designing complex electronic chips, avoiding the need to produce a prototype, as shown in the image in Figure 5.11, produced by the Altium Designer program. A 3D model of the PCB, components and casing can be produced directly in an electronic CAD tool to identify collisions between elements and carry out qualitative observations of the capacity of the designed system to allow an air flow to circulate through the casing. Moreover, it is possible to export these models to simulation tools in order to carry out finite element simulations (thermics, fluid mechanics, electromagnetic compatibility, etc.).

Appendix

Formulas for Electrical Engineering and Electromagnetism

A.1. Sinusoidal quantities

A.1.1. *Scalar signals*

A.1.1.1. *Definitions*

Sinusoidal waveforms are extremely widespread in electrical engineering, both for voltages and for currents. In this case, we will consider a generic signal of the form:

$$x\left(t\right) = X_{\max} \cos\left(\omega t - \varphi\right) \qquad \text{[A.1]}$$

This real signal is associated with an equivalent complex signal:

$$\underline{x}\left(t\right) = X_{\max}.\mathrm{e}^{j\left(\omega t - \varphi\right)} \qquad \text{[A.2]}$$

This vector may be represented in the complex plane. We obtain a circular trajectory of radius $X_{\max}$ with a vector rotating at a constant speed ω in a counterclockwise direction. This representation (which is widespread in electrical engineering) is known as a Fresnel diagram (or, more simply, a vector diagram).

REMARK A.1.– Derivation and integration calculations are greatly simplified in the complex plane, as they are replaced, respectively, by multiplying or dividing by $j\omega$. To return to the real domain, we simply take the real part of the corresponding complex signal: $x(t) = \mathfrak{Re}\left[\underline{x}\,(t)\right]$.

The rotating component $\mathrm{e}^{j\omega t}$ of the complex vectors is meaningless when studying linear circuits; the *amplitudes* and the relative phases between the different quantities under study are the only important elements. Note that an absolute phase for a sinusoidal value would be meaningless; the choice of a reference value of the form $X_{\mathrm{ref}}.\cos(\omega.t)$, associated with the vector $X_{\mathrm{ref}}.\mathrm{e}^{j\omega t}$, is purely arbitrary.

Complex vectors are also often represented (in the literature) using the RMS value of the real value in question as the modulus, and not the real amplitude.

A.1.1.2. *Trigonometric formulas*

When making calculations using complex values, we need Euler's formulas:

$$\begin{cases} \cos\theta = \frac{\mathrm{e}^{j\theta}+\mathrm{e}^{-j\theta}}{2} \\ \sin\theta = \frac{\mathrm{e}^{j\theta}-\mathrm{e}^{-j\theta}}{2j} \end{cases} \qquad [\mathrm{A.3}]$$

These two formulas can be used to give the four basic trigonometric formulas used in electrical engineering:

$$\begin{cases} \cos\left(a+b\right) = \cos a \cos b - \sin a \sin b \\ \cos\left(a-b\right) = \cos a \cos b + \sin a \sin b \\ \sin\left(a+b\right) = \sin a \cos b + \cos a \sin b \\ \sin\left(a-b\right) = \sin a \cos b - \cos a \sin b \end{cases} \qquad [\mathrm{A.4}]$$

These four equations allow us to establish four further equations:

$$\cos a \cos b = \frac{1}{2}\left(\cos\left(a+b\right) + \cos\left(a-b\right)\right) \qquad [\mathrm{A.5}]$$

$$\begin{cases} \cos a \cos b = \frac{1}{2}\left(\cos\left(a+b\right)+\cos\left(a-b\right)\right) \\ \sin a \sin b = \frac{1}{2}\left(\cos\left(a-b\right)-\cos\left(a-b\right)\right) \\ \sin a \cos b = \frac{1}{2}\left(\sin\left(a+b\right)+\sin\left(a-b\right)\right) \\ \cos a \sin b = \frac{1}{2}\left(\sin\left(a+b\right)-\sin\left(a-b\right)\right) \end{cases} \qquad \text{[A.6]}$$

A.1.2. *Vector signals (three-phase context)*

A.1.2.1. *Reference frame* (a, b, c)

Three-phase systems are very much common in electrical engineering, particularly balanced three-phase systems. A vector $(\mathbf{x}_3) = (x_a, x_b, x_c)^t$ with three balanced components is therefore expressed as:

$$(\mathbf{x}_3) = X_{\max} \begin{pmatrix} \cos\theta \\ \cos\left(\theta - \frac{2\pi}{3}\right) \\ \cos\left(\theta + \frac{2\pi}{3}\right) \end{pmatrix} \text{ where } \theta = \omega.t + \phi_0 \qquad \text{[A.7]}$$

in the case of a direct system, or:

$$(\mathbf{x}_3) = X_{\max} \begin{pmatrix} \cos\theta \\ \cos\left(\theta + \frac{2\pi}{3}\right) \\ \cos\left(\theta - \frac{2\pi}{3}\right) \end{pmatrix} \text{ where } \theta = \omega.t + \phi_0 \qquad \text{[A.8]}$$

in the inverse case.

DEFINITION A.1.– A balanced three-phase system is thus made up of three sinusoids of the same amplitude and same frequency, with a phase deviation of $\frac{2\pi}{3}$.

A direct three-phase system is characterized by the fact that, taking phase 1 as a reference point (i.e. first component), the second component has a delay of 120° (in a balanced situation) and the third component presents a delay of 120° in relation to the second component.

An inverse three-phase system is characterized by the fact that, taking phase 1 as a reference point (i.e. first component), the third component has a delay of 120° (in a balanced situation) and the second component presents a delay of 120° in relation to the third component. A direct system can be converted into an inverse system (and vice versa) by permutations of two components.

A.1.2.2. *Three-phase to two-phase transformation* (α, β)

It is important to note that a balanced three-phase system (whether direct or inverse) presents an important property in that the sum of the components is null:

$$x_a + x_b + x_c = 0 \qquad\qquad [\text{A.9}]$$

This sum is classically referred to as the zero sequence component (denoted as x_0). A balanced three-phase system is therefore not linearly independent in that, given two of the components, we may calculate the value of the third component. It is therefore possible to propose a three-phase to two-phase transformation without any information loss. The simplest transformation, known as the Clarke (abc-to-$\alpha\beta$) transformation, allows us to associate an initial vector $(\mathbf{x}_3) = (x_a, x_b, x_c)^t$ with an equivalent two-phase vector $(\mathbf{x}_{\alpha\beta}) = (\mathbf{x}_2) = (x_\alpha, x_\beta)^t$ using components of the same amplitude as those in the initial vector. This operation introduces the Clarke matrix C_{32}:

$$X_{\max} \begin{pmatrix} \cos\theta \\ \cos\left(\theta + \frac{2\pi}{3}\right) \\ \cos\left(\theta - \frac{2\pi}{3}\right) \end{pmatrix} = X_{\max} \cdot \underbrace{\begin{pmatrix} 1 & 0 \\ -1/2 & \sqrt{3}/2 \\ -1/2 & -\sqrt{3}/2 \end{pmatrix}}_{C_{32}} \cdot \begin{pmatrix} \cos\theta \\ \sin\theta \end{pmatrix} \qquad [\text{A.10}]$$

This gives the following direct transformation:

$$(\mathbf{x}_3) \triangleq C_{32} \cdot (\mathbf{x}_2) \qquad\qquad [\text{A.11}]$$

The Clarke transformation may be extended by taking account of the zero sequence component x_0, presented in [A.9]:

$$(\mathbf{x_3}) \triangleq C_{32}.(\mathbf{x_2}) + C_{31}.x_0 \qquad [\text{A.12}]$$

with:

$$C_{31} = \begin{pmatrix} 1 \\ 1 \\ 1 \end{pmatrix} \qquad [\text{A.13}]$$

Noting certain properties of matrices C_{32} and C_{31}:

$$C_{32}^t C_{32} = \frac{3}{2} \begin{pmatrix} 1 & 0 \\ 0 & 1 \end{pmatrix} \; ; \; C_{31}^t C_{31} = 3$$
$$C_{32}^t C_{31} = \begin{pmatrix} 0 \\ 0 \end{pmatrix} \; ; \; C_{31}^t C_{32} = \begin{pmatrix} 0 & 0 \end{pmatrix} \qquad [\text{A.14}]$$

we can establish the inverse transformation:

$$(\mathbf{x_2}) \triangleq \frac{2}{3} C_{32}^t.(\mathbf{x_3}) \qquad [\text{A.15}]$$

and:

$$x_0 \triangleq \frac{1}{3} C_{31}^t.(\mathbf{x_3}) \qquad [\text{A.16}]$$

A.1.2.3. *Concordia variant*

A second three-phase to two-phase transformation is also widely used in the literature, with properties similar to those of the Clarke transformation. This variation does not retain the amplitudes of the transformed values, but allows us to retain powers. This operation is known as the Concordia transformation and is based on two matrices, denoted T_{32} and T_{31}, deduced from C_{32} and C_{31}:

$$T_{32} = \sqrt{\frac{2}{3}} C_{32} \; ; \; T_{31} = \frac{1}{\sqrt{3}} C_{31} \qquad [\text{A.17}]$$

The properties of these matrices are deduced from those established in [A.14]:

$$T_{32}^t T_{32} = \begin{pmatrix} 1 & 0 \\ 0 & 1 \end{pmatrix} \; ; \; T_{31}^t T_{31} = 1$$

$$T_{32}^t T_{31} = \begin{pmatrix} 0 \\ 0 \end{pmatrix} \; ; \; T_{31}^t T_{32} = \begin{pmatrix} 0 & 0 \end{pmatrix}$$

[A.18]

This produces a direct transformation of the form:

$$(\mathbf{x}_3) \triangleq T_{32} . (\mathbf{x}_2) + T_{31} . x_0$$

[A.19]

with the following inverse transformation:

$$(\mathbf{x}_2) \triangleq T_{32}^t . (\mathbf{x}_3)$$

[A.20]

and:

$$x_0 \triangleq T_{31}^t . (\mathbf{x}_3)$$

[A.21]

A.1.2.4. *Park transformation*

The Park (abc-to-dq) transformation consists of associating the Clarke (or Concordia) transformation with a rotation in the two-phase reference plane (α, β) onto a reference frame (d, q). This operation is carried out using the rotation matrix $P(\theta)$, defined as:

$$P(\theta) = \begin{pmatrix} \cos\theta & -\sin\theta \\ \sin\theta & \cos\theta \end{pmatrix}$$

[A.22]

Thus, if we associate a vector $(\mathbf{x}_{dq}) = (x_d, x_q)^t$ with the initial two-phase vector $(\mathbf{x}_{\alpha\beta}) = (\mathbf{x}_2)$ (obtained from a Clarke or Concordia transformation), we obtain the following relationship:

$$(\mathbf{x}_{\alpha\beta}) = (\mathbf{x}_2) \triangleq P(\theta) . (\mathbf{x}_{dq})$$

[A.23]

The choice of a frame of reference involves the definition of angle θ, selected arbitrarily. Generally, the chosen reference frame is synchronous with the rotating values (sinusoidal components with an angular frequency ω), but this is not obligatory.

The following (non-exhaustive) list shows a number of properties of matrix $P(\theta)$:

$$P(0) = \begin{pmatrix} 1 & 0 \\ 0 & 1 \end{pmatrix} = \mathbb{I}_2 \; ; \; P\left(\tfrac{\pi}{2}\right) = \begin{pmatrix} 0 & -1 \\ 1 & 0 \end{pmatrix} \quad \text{[A.24]}$$
$$= \mathbb{J}_2 \text{ such that } \mathbb{J}_2 = -\mathbb{I}_2$$

$$P(\alpha + \beta) = P(\beta + \alpha) = P(\alpha).P(\beta) = P(\beta).P(\alpha) \quad \text{[A.25]}$$

$$P(\alpha)^{-1} = P(\alpha)^t = P(-\alpha) \quad \text{[A.26]}$$

$$\tfrac{d}{dt}[P(\alpha)] = \tfrac{d\alpha}{dt} \cdot P\left(\alpha + \tfrac{\pi}{2}\right) = \tfrac{d\alpha}{dt} \cdot P(\alpha) \cdot P\left(\tfrac{\pi}{2}\right)$$
$$= \mathbb{J}_2 \tfrac{d\alpha}{dt} \cdot P(\alpha) \quad \text{[A.27]}$$

A.1.2.5. *Phasers or complex vectors*

The matrix formalism of the Clarke, Concordia and Park transformations may be replaced by an equivalent complex representation. Evidently, a rotation of the frame of reference by angle θ may be obtained by using a complex coefficient $e^{j\theta}$ as easily as with a rotation matrix $P(\theta)$. To this end, we use a "phaser" $\underline{x}_s$ defined in a stationary frame of reference:

$$\underline{x}_s = x_\alpha + j.x_\beta \quad \text{[A.28]}$$

The phaser is also defined in a rotating frame ($\underline{x}_r$):

$$\underline{x}_r = x_d + j.x_q \quad \text{[A.29]}$$

Note that these complex representations may be obtained using matrix transformations. The real transformations seen in the previous sections each have an equivalent complex transformation, as shown in Table A.1.

Real transformation	Complex transformation
Clarke	Fortescue
Concordia	Lyon
Park	Ku

Table A.1. *Correspondence between real and complex transformations (names)*

A.2. General characteristics of signals in electrical engineering

This section presents the formulas used for calculating the *general characteristics of periodic signals* traditionally encountered in electrical engineering. However, it does not cover formulas related to spectral analysis, which are covered in the Appendices 2 of Volumes 2 and 4 of this series.

In this section, we will therefore cover the formulas used to calculate the average and RMS values of given quantities, applied to two widespread signal types: sinusoids and the asymmetric square signal of duty ratio α.

A.2.1. *Average value*

A.2.1.1. *General definition*

The average value $\langle x \rangle$ of a T-periodic signal $x(t)$ is defined generally by the integral:

$$\langle x \rangle = \frac{1}{T} \int_0^T x(t).dt \qquad [\text{A.30}]$$

REMARK A.2.– In this case, the integration limits are chosen arbitrarily. Only the interval between the two limits is important, and it must be equal to T.

A.2.1.2. *Sinusoids*

In the case of sinusoids, we evidently obtain an average value of zero.

A.2.1.3. *Asymmetric square*

The T-periodic asymmetric square $x(t)$ studied here has a certain value X_0 during a period αT, then 0 for the rest of the period. We can therefore write the average value $\langle x \rangle$ directly:

$$\langle x \rangle = \frac{1}{T} \int_0^T x(t).dt = \frac{1}{T} \int_0^{\alpha T} X_0.dt = \alpha.X_0 \qquad \text{[A.31]}$$

A.2.2. *RMS value*

A.2.2.1. *General definition*

The RMS value X_{rms} of a T-periodic signal $x(t)$ is defined generally by the integral:

$$X_{\text{rms}} = \sqrt{\frac{1}{T} \int_0^T x^2(t).dt} \qquad \text{[A.32]}$$

REMARK A.3.– When calculating the average value, the integration limits are chosen arbitrarily. Only the interval between the two limits is important, and it must be equal to T.

A.2.2.2. *Sinusoids*

For a sinusoid of amplitude X_{max}, the RMS value is $X_{\text{rms}} = \frac{X_{\text{max}}}{\sqrt{2}}$.

A.2.2.3. *Asymmetric square*

The T-periodic asymmetric square $x(t)$ defined in section A.2.1 presents an RMS value expressed as:

$$X_{\text{rms}} = \sqrt{\frac{1}{T} \int_0^{\alpha T} X_0^2.dt} = \sqrt{\alpha}.X_0 \qquad \text{[A.33]}$$

A.3. Energy and power

A.3.1. *Energy*

In mechanics, energy is obtained by the operation of a force over a certain distance. In electrical engineering, this term corresponds to the movement of a charge following a variation in electrical potential. In particle physics, a unit known as an electron-volt (eV) is used for energy values at the atomic level. The energy formulas used in power electronics (expressed in Joules (J)) correspond to the energy stored in an inductance or a capacitor.

In an inductance, the energy E_L (magnetic energy) depends on the current I and the inductance L:

$$E_L = \frac{1}{2}LI^2 \qquad \text{[A.34]}$$

For a capacitor, the energy E_C (electrostatic energy) depends on the voltage V and the capacitance C:

$$E_C = \frac{1}{2}CV^2 \qquad \text{[A.35]}$$

A.3.2. *Instantaneous power*

The instantaneous power $p(t)$ given – or provided to – the dipole is linked, according to the passive sign convention (PSC), to the voltage $v(t)$ at its terminals and the current $i(t)$ passing through it as follows:

$$p(t) = v(t).i(t) \qquad \text{[A.36]}$$

This power is defined in watts (W). It is linked to the energy consumed E (in J) between two instants t_1 and t_2 by the following integral:

$$E = \int_{t_1}^{t_2} p(t).dt \qquad \text{[A.37]}$$

The instantaneous power $p(t)$ is connected to the variation in energy $e(t)$ which can also be defined (up to an additive constant) as a function of time. In this case, we obtain:

$$p(t) = \frac{de(t)}{dt} \qquad \text{[A.38]}$$

A.3.3. Average power

As for any T-periodic signal, the average power P is obtained using the following formula:

$$P = \frac{1}{T} \int_0^T p(t).dt = \frac{1}{T} \int_0^T v(t).i(t).dt \qquad \text{[A.39]}$$

In the case of a resistive charge R, we can establish the following relationship (Ohm's law):

$$v(t) = R.i(t) \qquad \text{[A.40]}$$

This allows us to formulate two possible expressions for this power:

$$P = \frac{R}{T} \int_0^T i^2(t).dt = R.I_{\text{rms}}^2 \qquad \text{[A.41]}$$

and:

$$P = \frac{1}{RT} \int_0^T v^2(t).dt = \frac{V_{\text{rms}}^2}{R} \qquad \text{[A.42]}$$

where V_{rms} and I_{rms} are the RMS values of the voltage and the current, respectively.

A.3.4. *Sinusoidal mode*

A.3.4.1. *Single phase*

In single phase sinusoidal operating mode, we can, generally speaking, consider a voltage $v(t)$ of the form:

$$v(t) = V_{\text{rms}}\sqrt{2}\cos(\omega t) \qquad [\text{A.43}]$$

as the phase reference, with a current, with a phase deviation angle φ (the lag in relation to the voltage), expressed as:

$$i(t) = I_{\text{rms}}\sqrt{2}\cos(\omega t - \varphi) \qquad [\text{A.44}]$$

Calculating the instantaneous power obtained using these two values, we obtain:

$$p(t) = V_{\text{rms}}I_{\text{rms}}\left(\cos(2\omega t - \varphi) + \cos\varphi\right) \qquad [\text{A.45}]$$

We thus obtain two terms:

– a constant term, which is, evidently, the average power, referred to in this context as active power;

– a variable term, with an angular frequency of 2ω, known as fluctuating power.

The first interesting result is, therefore, the expression of the average (active) power P:

$$P = V_{\text{rms}}I_{\text{rms}}\cos\varphi \qquad [\text{A.46}]$$

In terms of voltage dimensioning (thickness of insulation) and current dimensioning (cross-section of conductors) of equipment, the real power value used for design purposes is known as the apparent power S , and is obtained by directly multiplying the RMS voltage value by the RMS current value:

$$S = V_{\text{rms}}I_{\text{rms}} \qquad [\text{A.47}]$$

To emphasize the "fictional" character of this power, it is not given in W, but in volt-amperes (VA).

In electrical engineering, we then use the notion of *reactive power* Q, which allows us to establish a connection between the active power P and the apparent power S. This is expressed as:

$$Q = V_{rms} I_{rms} \sin \varphi \qquad \text{[A.48]}$$

The connection between P, Q and S is thus:

$$S^2 = P^2 + Q^2 \qquad \text{[A.49]}$$

As in the case of apparent power, this power value is fictional; it is not measured in W, or in VA, but rather in volt ampere reactive (VAR).

REMARK A.4.– Equation [A.49] is only valid if the voltage *and* the current are sinusoidal. In non-sinusoidal mode, we introduce an additional power, denoted D, known as the deformed power. This is used to establish a new equation as follows:

$$S^2 = P^2 + Q^2 + D^2 \qquad \text{[A.50]}$$

The instantaneous power is always positive (respectively, negative) when $\varphi = 0°$ (respectively, $\varphi = 180°$), but if φ takes a different value, $p(t)$ cancels out, changing the sign. In these conditions, the direction of transfer of electronic energy between the source and the load is reversed.

A.3.4.2. *Three phase*

In a three-phase context, using the "voltage" vector (v_3) as a point of reference, and more specifically as the first component, we take (based on the hypothesis of a direct balanced system):

$$(\mathbf{v}_3) = V_{\mathrm{rms}}\sqrt{2} \begin{pmatrix} \cos(\omega t) \\ \cos\left(\omega t - \frac{2\pi}{3}\right) \\ \cos\left(\omega t + \frac{2\pi}{3}\right) \end{pmatrix} \qquad [\text{A.51}]$$

From this, we deduce the "current" vector $(\mathbf{i}_3)$, with a lag in each component when compared to the corresponding components in $(\mathbf{v}_3)$:

$$(\mathbf{i}_3) = I_{\mathrm{rms}}\sqrt{2} \begin{pmatrix} \cos(\omega t - \varphi) \\ \cos\left(\omega t - \frac{2\pi}{3} - \varphi\right) \\ \cos\left(\omega t + \frac{2\pi}{3} - \varphi\right) \end{pmatrix} \qquad [\text{A.52}]$$

A matrix formalism may be used to obtain the expression of the instantaneous power $p(t)$:

$$p(t) = (\mathbf{v}_3)^t . (\mathbf{i}_3) \qquad [\text{A.53}]$$

In this case, the Park factorization of the "voltage and current" vectors is particularly effective:

$$(\mathbf{v}_3) = V_{\mathrm{rms}}\sqrt{2}.C_{32} \begin{pmatrix} \cos(\omega t) \\ \sin(\omega t) \end{pmatrix}$$

$$= V_{\mathrm{rms}}\sqrt{2}.C_{32}.P(\omega t) . \begin{pmatrix} 1 \\ 0 \end{pmatrix} \qquad [\text{A.54}]$$

$$(\mathbf{i}_3) = I_{\mathrm{rms}}\sqrt{2}.C_{32} \begin{pmatrix} \cos(\omega t - \varphi) \\ \sin(\omega t - \varphi) \end{pmatrix}$$

$$= I_{\mathrm{rms}}\sqrt{2}.C_{32}.P(\omega t - \varphi) . \begin{pmatrix} 1 \\ 0 \end{pmatrix} \qquad [\text{A.55}]$$

Hence:

$$p(t) = 2V_{\text{rms}}.I_{\text{rms}} \begin{pmatrix} 1 & 0 \end{pmatrix}.P\left(-\omega t\right).C_{32}^t.C_{32}.P\left(\omega t - \varphi\right).\begin{pmatrix} 1 \\ 0 \end{pmatrix} \text{[A.56]}$$

After simplification, this gives us:

$$p(t) = 3V_{\text{rms}}.I_{\text{rms}}\cos\varphi \qquad \text{[A.57]}$$

Note that we obtain the instantaneous power, and not an average value. This highlights a notable property of three-phase systems: there is no globally fluctuating power in this configuration.

The active power P is therefore defined as follows:

$$P = p(t) = 3V_{\text{rms}}.I_{\text{rms}}\cos\varphi \qquad \text{[A.58]}$$

The notions of reactive power Q and apparent power S are also used in three-phase contexts, with the following expressions:

$$\begin{cases} Q = 3V_{\text{rms}}.I_{\text{rms}}\sin\varphi \\ S = 3V_{\text{rms}}.I_{\text{rms}} \end{cases} \qquad \text{[A.59]}$$

Relationship [A.49] is therefore still valid in a three-phase context:

$$S^2 = P^2 + Q^2 \qquad \text{[A.60]}$$

Note that variants exist, notably where the notion of line-to-line voltage is used. Voltage V_{rms} is the RMS *line-to-neutral voltage* (i.e. between the phase and the neutral); the neutral is not always accessible, so the notion of line-to-line voltage is often preferred , with an RMS voltage, denoted U_{rms}. In the case of a balanced three-phase system, the relationship between the RMS line-to-neutral and line-to-line voltage is:

$$U_{\mathrm{rms}} = \sqrt{3}V_{\mathrm{rms}} \qquad\qquad \text{[A.61]}$$

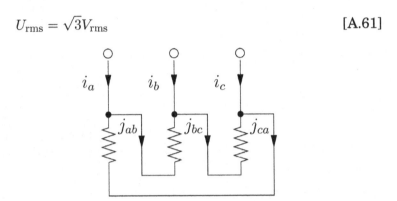

Figure A.1. *Line and branch currents for a triangular connection*

A second point, which may lead to a different formulation of expression [A.58], is concerned with currents. Generally speaking, we always have access to *line currents*, and thus to the RMS value I_{rms}. A second type of current can appear when using a load with a triangle connection (see Figure A.1): this branch current presents an RMS value J_{rms} with the following expression as a function of I_{rms}:

$$J_{\mathrm{rms}} = \frac{I_{\mathrm{rms}}}{\sqrt{3}} \qquad\qquad \text{[A.62]}$$

A.4. Mathematics for electromagnetism

A.4.1. *The Green–Ostrogradsky theorem*

The Green–Ostrogradsky theorem (also known as the flux–divergence theorem) establishes a connection between the integral of the divergence of a field with vector E in a volume Ω and the integral of the flux of E on the closed surface $\partial\Omega$ delimiting the volume Ω:

$$\iiint\limits_{\Omega} \mathrm{div}\mathbf{E}.d\omega = \oiint\limits_{\partial\Omega} \mathbf{E} \cdot ds \qquad [\text{A.63}]$$

where $d\omega$ is a volume element, while ds is a normal vector[1] with a surface element (infinitesimal) ds of the complete surface $\partial\Omega$.

A.4.2. Stokes–Ampère theorem

The Stokes–Ampère theorem establishes a connection between the integral of the rotational flux of a field with vector H on a surface Σ and the integral of the circulation of H along the closed contour $\partial\Sigma$ delimiting surface Σ:

$$\iint\limits_{\Sigma} \mathrm{rot}\,\mathbf{H} \cdot ds = \oint\limits_{\partial\Sigma} \mathbf{H} \cdot dl \qquad [\text{A.64}]$$

where ds is a normal vector[2] with a surface element (infinitesimal) ds of the complete surface Σ. Element dl is a norm vector dl carried by the closed contour $\partial\Sigma$.

A.4.3. Differential and referential operators

The definition of the differential operators used in electromagnetism (primarily grad, div and rot) is dependent on the chosen frame of reference. Using the Cartesian

---

1 Oriented toward the outside of volume Ω.
2 Oriented in accordance with the right-hand rule as a function of the choice of orientation of contour $\partial\Sigma$.

coordinate system, the nabla operator (vector), ∇, allows us to easily write these operators as:

$$\nabla = \begin{pmatrix} \frac{\partial}{\partial x} \\ \frac{\partial}{\partial y} \\ \frac{\partial}{\partial z} \end{pmatrix} \qquad\qquad [A.65]$$

and we know that:

$$\begin{cases} \operatorname{grad} V = \nabla V \\ \operatorname{div} \mathbf{E} = \nabla \cdot \mathbf{E} \\ \operatorname{rot} \mathbf{H} = \nabla \times \mathbf{H} \end{cases} \qquad\qquad [A.66]$$

where the symbol "$\cdot$" is the scalar product and "$\times$" is the vector product.

If we want to write these operators using spherical or cylindrical coordinates, the ∇ operator is no longer suitable; in these cases, it is better to use intrinsic definitions (which are independent of the chosen frame of reference). For the gradient, we have:

$$dV = (\operatorname{grad} V) \cdot d\mathbf{r} \qquad\qquad [A.67]$$

where dV is the exact total differential of V and $d\mathbf{r}$ is an infinitesimal shift (vector) away from the considered point in the space (defined by vector r from the origin of the reference frame).

For the "divergence" and "rotational" operators, we simply use the two theorems presented in sections A.4.1 and A.4.2. First, we obtain:

$$d\phi = \operatorname{div} \mathbf{E}.d\omega \qquad\qquad [A.68]$$

where $d\phi$ is the flux of $\mathbf{E}$ across the surface of the volume $d\omega$ under consideration.

We can then write:

$$dC = \text{rot}\,\mathbf{H} \cdot \text{n}.dS \qquad\qquad [\text{A.69}]$$

where dC is the circulation of field H along a closed contour enclosing a surface dS, and with an orientation allowing us to define a normal (unitary) vector n (in accordance with the right-hand rule).

Bibliography

[APP 02] APPEL W., *Mathématiques pour la physique et les physiciens*, H & K, 2002.

[BAS 09] BASDEVANT J.-L., *Les mathématiques de la physique quantique*, Vuibert, 2009.

[BEC 00] BECH M.M., Analysis of random pulse-width modulation techniques for power electronic converters, Doctoral Thesis, Aalborg University, 2000.

[BEN 33] BENNETT W.R., "New results in the calculation of modulation productions", *Bell System Technical Journal*, vol. 12, no. 4, pp. 238–243, 1933.

[BIE 08] BIERHOFF M., FUCHS F.W., "DC link harmonics of three phase voltage source converters influenced by the pulsewidth modulation strategy – an analysis", *IEEE Transactions on Industrial Electronics*, vol. 55, no. 5, pp. 2085–2092, May 2008.

[BLA 53] BLACK H.S., *Modulation Theories*, Van Nostrand, New York, 1953.

[BOW 08] BOWICK C., BLYLER J., AJLUNI C., *RF Circuit Design*, Newnes/Elsevier, 2008.

[BRÉ 05] BRÉHAUT S., Modélisation et optimisation des performances CEM d un convertisseur AC/DC d'une puissance de 600 W, Doctoral Thesis, François Rabelais University, Tours, 2005.

[BÜH 91] BÜHLER H., *Convertisseurs Statiques*, Presses Polytechniques et Universitaires Romandes, 1991.

[COR 13] CORNELL DUBILIER, *Aluminum Electrolytic Capacitor Application Guide*, Cornell Dubilier, 2013. Available at http://www.cde.com/fliptest/alum/alum.html.

[CHA 05] CHAROY A., *Compatibilité électromagnétique*, Dunod, 2005.

[CHE 09] CHENAND L., PENG F.Z., "Closed-loop gate drive for high power IGBTs", *Proceedings of the IEEE APEC*, pp. 1331–1337, 2009.

[CHE 99] CHERON Y., *La commutation douce*, Tec & Doc Lavoisier, 1999.

[COH 00] COHEN DE LARA M., D'ANDRÉA-NOVEL B., Cours d'automatique – commande linéaire des systèmes dynamiques, Mines Press, 2000.

[COC 02] COCQUERELLE J.-L., PASQUIER C., *Rayonnement électromagnétique des convertisseurs à découpage*, EDP Sciences, 2002.

[COS 05] COSTA F., MAGNON D., "Graphical analysis of the spectra of EMI sources in power electronics", *IEEE Transactions on Power Electronics*, vol. 20, no. 6, pp. 1491–1498, November 2005.

[COS 13] COSTA F., GAUTIER C., LABOURÉ E., *et al.*, *La compatibilité électromagnétique en électronique de puissance, Principes et cas d'études*, Hermès-Lavoisier, Paris, 2013.

[DEG 01] DEGRANGE B., Introduction à la physique quantique, Mines Press, 2001.

[FER 02] FERRIEUX J.-P., FOREST F., *Alimentations à découpages et convertisseurs à résonance*, 3rd ed., Dunod, 2002.

[FEY 99] FEYNMAN R., LEIGHTON R.B., SANDS M., *Cours de physique de Feynman, Electromagnétisme*, vol. 1–2, Dunod, 1999.

[FOC 98] FOCH H., FOREST F., MEYNARD T., Onduleurs de tension – Structures, Principes, Applications, Technical Engineer, Traité Génie Electrique, 1998.

[FOC 94] FOCH H., CHÉRON Y., "Convertisseur de type Forward", *Techniques de l'Ingénieur*, vol. 5, pp. 3167.1–3167.10, 1994.

[FRI 94] FRICKEY D.A., "Conversions between S, Z, Y, h, ABCD, and T parameters which are valid for complex source and load impedances", *IEEE Transactions on Microwave Theory and Techniques*, vol. 42, no. 2, 1994.

[GHA 03] GHAUSI M., LAKER K., *Modern Filter Design: Active RC and Switched Capacitor*, Noble Publishing, 2003.

[GIB 07] GIBSON W.C., *The Method of Moments in Electromagnetics*, Chapman & Hall/CRC, 2007.

[HAV 98] HAVA A.M., KERKMAN R., LIPO T., "A high performance generalized discontinuous PWM algorithm", *IEEE Transactions on Industry Applications*, vol. 34, no. 5, pp. 1059–1071, 1998.

[HAV 99] HAVA A.M., LIPO T.A., KERKMAN R.J., "Simple analytical and graphical methods for carrier-based PWM-VSI drives", *IEEE Transactions on Power Electronics*, vol. 14, no. 1, pp. 49–61, 1999.

[HOB 05] HOBRAICHE J., Comparaison des stratégies de modulation à largeur d'impulsions triphasées – Application à l'alterno-démarreur, Doctoral Thesis, UTC, Compiègne, 2005.

[HOL 83] HOLTZ J., STADTFELD S., "A predictive controller for a stator current vector of AC-machines fed from a switched voltage source", *Proceedings of the International Power Electronics Conference (IPEC)*, vol. 2, pp. 1665–1675, Tokyo, 1983.

[IEE 12] IEEE STANDARDS ASSOCIATION, *IEEE Electromagnetic Compatibility Standards Collection: VuSpec*™, CD-ROM, IEEE Standards Association, 2012.

[KEM 12] KEMET, *Electrolytic Capacitors*, Documentation technique, FF3304 06/09, 2012. Available at: www.kemet.com.

[KOL 91] KOLAR J.W., ERLT H., ZACH F.C., "Influence of the modulation method on the conduction and switching losses of a PWM converter system", *IEEE Transactions on Industry Applications*, vol. 27, no. 6, pp. 399–403, 1991.

[KOL 93] KOLAR J.W., Vorrichtung und Verfahren zur Umformung von Drehstrom in Gleichstrom, IXYS Semiconductor GmbH, Patent no. EP0660498 A2, 23 December 1993.

[LAN 09] LANFRANCHI V., PATIN N., DÉPERNET D., "MLI précalculées et optimisées", Chapter 4 in MONMASSON E., (ed.), *Commande rapprochée de convertisseurs*, Hermès-Lavoisier, 2009.

[LEF 02] LEFEBVRE S., MULTON B., "Commande des semi-conducteurs de puissance: principes", *Techniques de l'Ingénieur*, vol. 5, pp. 3231.1–3231.23, 2002.

[LES 97] LESBROUSSARD C., Etude d'une stratégie de modulation de largeur d'impulsions pour un onduleur de tension triphasé à deux ou trois niveaux: la Modulation Delta Sigma Vectorielle, Doctoral Thesis, UTC, 1997.

[LIN 10] LINDER A., KANCHAN R., KENNEL R., *et al.*, *Model-Based Predictive Control of Electrical Drives*, Cuvillier Verlag, Göttingen, 2010.

[LUM 00] LUMBROSO H., *Ondes électromagnétiques dans le vide et les conducteurs, 70 problèmes résolus*, 2nd ed., Dunod, 2000.

[MAT 09] MATHIEU H., FANET H., *Physique des semiconducteurs et des composants*, 6th ed., Dunod, 2009.

[MEY 93] MEYNARD T., FOCH H., "Imbricated cells multilevel voltage-source inverters for high-voltage applications", *EPE Journal*, vol. 3, pp. 99–106, June 1993.

[MIC 05] MICROCHIP, Sinusoidal Control of PMSM Motors with dsPIC30F3010, AN1017, 2005. Available at http://www.microchip.com/downloads/en/AppNotes/01017A.pdf.

[MID 77] MIDDLEBROOK R.D., CÙK S., "A general unified approach to modeling switching converter power stages", *International Journal of Electronics*, vol. 42, no. 6, pp. 521–550, 1977.

[MOH 95] MOHAN N., UNDELAND T.M., ROBBINS W.P., *Power Electronics – Converters, Applications and Design*, 2nd ed., Wiley, 1995.

[MON 97] MONMASSON E., FAUCHER J., "Projet pédagogique autour de la MLI vectorielle", *Review*, vol. 3EI, no. 8, pp. 22–36, 1997.

[MON 09] MONMASSON E., (ed.), *Commande rapprochée de convertisseurs statiques 1, Modulation de largeur d'impulsion*, Hermès-Lavoisier, Paris, 2009.

[MON 11] MONMASSON E., (ed.), *Power Electronic Converters: PWM Strategies and Current Control Techniques*, ISTE, London and John Wiley & Sons, New York, 2011.

[MOR 07] MOREL F., Commandes directes appliquées à une machine synchrone à aimants permanents alimentée par un onduleur triphasé à deux niveaux ou par un convertisseur matriciel triphasé, Doctoral Thesis, INSA Lyon, December 2007.

[MOY 98] MOYNIHAN J.F., EGAN M.G., MURPHY J.M.D., "Theoretical spectra of space vector modulated waveforms", *IEEE Proceedings Electrical Power Applications*, vol. 145, pp. 17–24, 1998.

[MUK 10] MUKHTAR A., *High Performance AC Drives*, Springer, 2010.

[NAR 06] NARAYANAN G., KRISHNAMURTHY H.K., ZHAO D., *et al.*, "Advanced bus-clamping PWM techniques based on space vector approach", *IEEE Transactions on Power Electronics*, vol. 21, no. 4, pp. 974–984, 2006.

[NAR 08] NARAYANAN G., RANGANATHAN V.T., ZHAO D., *et al.*, "Space vector based hybrid PWM techniques for reduced current ripple", *IEEE Transactions on Industrial Electronics*, vol. 55, no. 4, pp. 1614–1627, 2008.

[NGU 11a] NGUYEN T.D., Etude de stratégies de modulation pour onduleur triphasé dédiées à la réduction des perturbations du bus continu en environnement embarqué, Doctoral Thesis, UTC, Compiègne, France 2011.

[NGU 11b] NGUYEN T.D., HOBRAICHE J., PATIN N., *et al.*, "A direct digital technique implementation of general discontinuous pulse width modulation strategy", *IEEE Transactions on Industrial Electronics*, vol. 58, no. 9, pp. 4445–4454, September 2011.

[OSW 11] OSWALD N., STARK B., HOLLIDAY D., *et al.*, "Analysis of shaped pulse transitions in power electronic switching waveforms for reduced EMI generation", *IEEE Transactions on Industry Applications*, vol. 47, no. 5, pp. 2154–2165, October–November 2011.

[PAT 15a] PATIN N., *Power Electronics Applied to Industrial Systems and Transports Volume 2*, ISTE Press, London and Elsevier, 2015.

[PAT 15b] PATIN N., *Power Electronics Applied to Industrial Systems and Transports Volume 3*, ISTE Press, London and Elsevier, 2015.

[PAT 15c] PATIN N., *Power Electronics Applied to Industrial Systems and Transports Volume 4*, ISTE Press, London and Elsevier, 2015.

[REB 98] REBY F., BAUSIERE R., SOHIER B., *et al.*, "Reduction of radiated and conducted emissions in power electronic circuits by the continuous derivative control method (CDCM)", *Proceedings of the 7th International Conference on Power Electronics and Variable Speed Drives*, pp. 158–162, 1998.

[REV 03] REVOL B., Modélisation et optimisation des performances CEM d'une association variateur de vitesse – machine asynchrone, Doctoral Thesis, Joseph Fourier University, Grenoble 1, Grenoble, 2003.

[ROM 86] ROMBAULT C., Séguier G., Bausière R., *L'électronique de puissance – Volume 2, La conversion AC-AC*, Tec & Doc, Lavoisier, 1986.

[ROU 04a] ROUDET J., CLAVEL E., GUICHON J.M., *et al.*, Modélisation PEEC des connexions dans les convertisseurs de puissance, *Techniques de l'Ingénieur*, vol. D30471, pp. 1–12, 2004.

[ROU 04b] ROUDET J., CLAVEL E., GUICHON J.M., Application de la méthode PEEC au cablage d'un onduleur triphasé, Technical de L'ingénieur, vol. D3072, pp. 1–10, 2004.

[ROU 04c] ROUSSEL J.-M., *Problèmes d'électronique de puissance*, Dunod, 2004.

[SCH 99] SCHELLMANNS A., Circuits équivalents pour transformateurs multienroulements. Application à la CEM conduite d'un convertisseur, INPG Doctoral Thesis, July 1999.

[SCH 01] DELABALLE J., La CEM: la compatibilité électromagnétique, Technical Manual no. 149, Schneider Electric, 2001.

[SÉG 11] SÉGUIER G., DELARUE P., LABRIQUE F., *Electronique de puissance*, Dunod, 2011.

[SHU 11] SHUKLA A., GHOSH A., JOSHI A., "Natural balancing of flying capacitor voltages in multicell inverter under PD carrier-based PWM", *IEEE Transactions on Power Electronics*, vol. 56, no. 6, pp. 1682–1693, June 2011.

[VAS 12] VASCAS F., IANNELI L., (eds.), *Dynamics and Control of Switched Electronic Systems*, Springer, 2012.

[VEN 07] VENET P., Amélioration de la sûreté de fonctionnement des dispositifs de stockage d'énergie, HDR Memory, Claude Bernard University – Lyon 1, Lyon, 2007.

[VIS 07] VISSER J.H., Active converter based on the Vienna rectifier topology interfacing a three-phase generator to a DC-bus, Master's Thesis, University of Pretoria, South Africa, 2007.

[VIS 12] VISSER H.J., Antenna Theory and Applications, Wiley, 2012.

[VOG 11] VOGELSBERGER M.A., WIESINGER T., ERTL H., "Lifecycle monitoring and voltage-managing unit for DC-link electrolytic capacitors in PWM converters", IEEE Transactions on Power Electronics, vol. 26, no. 2, pp. 493–503, February 2011.

[WEE 06] WEENS Y., Modélisation des câbles d'énergie soumis aux contraintes générées par les convertisseurs électroniques de puissance, Doctoral Thesis, USTL, Lille, 2006.

[WHE 04] WHEELER P.W., CLARE J.C., EMPRINGHAM L., et al., "Matrix converters", IEEE Industry Applications Magazine, vol. 10, no. 1, pp. 59–65, January–February 2004.

[WIL 99] WILLIAMS T., Compatibilité électromagnétique – De la conception à l'homologation, Publitronic, 1999.

[XAP 05] ALEXANDER M., Power distribution system (PDS) design: using bypass/decoupling capacitors, Application no. XAPP623, February 2005. Available at www.xilinx.com.

[YUA 00] YUAN X., BARBI I., "Fundamentals of a new diode clamping multilevel inverter", IEEE Transactions on Power Electronics, vol. 15, no. 4, pp. 711–718, July 2000.

[ZHA 10] ZHAO D., HARI V.S.S.P.K., NARAYANAN G., et al., "Space-vector-based hybrid pulsewidth modulation techniques for reduced harmonic distortion and switching loss", IEEE Transactions on Power Electronics, vol. 25, no. 3, pp. 760–774, 2010.

Index

Printed in the United States
By Bookmasters